Field G
of North America

Field Guide to Rivers of North America

Edited by

Arthur C. Benke
Department of Biological Sciences
University of Alabama
Tuscaloosa, Alabama

Colbert E. Cushing
Streamside Programs
Estes Park, Colorado

AMSTERDAM • BOSTON • HEIDELBERG • LONDON
NEW YORK • OXFORD • PARIS • SAN DIEGO
SAN FRANCISCO • SINGAPORE • SYDNEY • TOKYO
Academic Press is an imprint of Elsevier

Academic Press is an imprint of Elsevier
30 Corporate Drive, Suite 400, Burlington, MA 01803, USA
525 B Street, Suite 1900, San Diego, California 92101-4495, USA
84 Theobald's Road, London WC1X 8RR, UK

Library of Congress Cataloging-in-Publication Data
A catalog record for this book is available from the Library of Congress

British Library Cataloguing-in-Publication Data
A catalogue record for this book is available from the British Library

ISBN: 978-0-12-375088-4

For information on all Academic Press publications
visit our Web site at: www.elsevierdirect.com

Printed in China
09 10 11 12   6 5 4 3 2 1

Contents

Preface

In 2005, we edited and Academic Press/Elsevier published a large book entitled *Rivers of North America*, the first major reference volume on this topic. As the continent's most important natural resource, we felt that synthesizing the knowledge on these vital natural ecosystems was essential if we are to appreciate their importance and manage them wisely. The book contains 22 chapters that present extensive environmental and cultural information on several rivers in a particular drainage basin or geographical region extending from southern Mexico to the Arctic. A unique feature of the book is that in addition to text, each chapter includes a series of single-page "snapshots" of several rivers in that particular basin/region. Each of these pages contains: (1) a color map of the river's basin with topographic features, major dams, major population centers, physiographic provinces, etc., and (2) summary information on selected biological, chemical, physical, and geomorphological characteristics, including species lists for major animals and plants.

We have received considerable positive comment on the value of these single-page summaries and have decided to republish them in a compact, paperback version that enables rapid reference and is easy to carry on trips. This is what you are holding now. We have selected 200 of the most important or regionally representative rivers in North America for inclusion. Each river is featured on a colorful, two-page spread that includes the original topographic map, and most of the original environmental information. In addition, a color photograph is included for every river. We hope you find *Field Guide to the Rivers of North America* a useful addition to your library and companion in your travels. This book will help you more fully understand and appreciate the rivers you encounter and about which you are interested in learning. For readers wanting more information on these rivers, we urge you to consult our 2005 volume.

Arthur C. Benke
Colbert E. Cushing

Acknowledgments

First and foremost, we want to acknowledge the contributions of chapter authors. This book would not have been possible without their first-hand knowledge. We sincerely thank them for their considerable efforts.

We are especially grateful to Emily McCloskey and Candice Janco of Elsevier, who helped guide us through the conversion of portions of *Rivers of North America* to *Field Guide to Rivers of North America.*

We especially want to thank those associated with the creation of the color topographic maps of all river basins used throughout the book. Craig Remington, Director of the University of Alabama Cartography Laboratory (Department of Geography) supervised the maps project. Angela Brink spent many hours creating the maps and Jennifer White was instrumental in making many subsequent revisions as the book progressed.

We thank the many individuals who contributed photographs that greatly enhanced the beauty and appeal of this *Guide*. We are especially grateful to Tim Palmer for his generosity in sharing so many beautiful photos from his collection, including 2 photos from his newest river photo book *Rivers of America* by publisher Harry N. Abrams. Tim has written eloquently about the flowing waters of the United States and the importance of their conservation in *America by Rivers, Lifelines: The Case for River Conservation, Endangered Rivers and the Conservation Movement, The Wild and Scenic Rivers of America, The Snake River: Window to the West,* and others. Tim's contributions have not only added to the attractiveness of the *Guide*, but have filled some very important gaps. We also thank Beth Maynor Young, another professional photographer who contributed several striking photographs for Chapter 4. Thanks also go to many other individuals (besides chapter authors and editors), agencies, and organizations that contributed one or more photographs: A. Belala, J. Benstead, D. Bicknell, C. Bourgeois, J. Boynton, C. S. Brehme, H. Burian, J. Burrows, C. Burton, B. Caldwell, T. Carter, M. Chapman, M. Conly, D. J. Cooper, D. D. Dauble, W. E. Doolittle, K. Ekelund, B. Emerson, J. M. Farrell, S. Fend, P. Filippelli, K. Freeman, A. Garza, M. Gautreau, W. Gibbins, Great Valley Museum, T. Grey, D. Hall, G. Harris, T. Harris, C. Hayne, M. Hernandez-Henriquez, C. Holland, R. Holmes, M. Keckhaver, E. Martyn, E. McKittrick, D. M. Merritt, NASA, NOAA, North Dakota Game and Fish Department, M. O'Malley, D. Moug, B. Oswald, R. Overman, G. Pitchford, J. Powers, D. Ramsey, J. Rathert, M. Rautio, L. Renan, J. Robinson,

K. Schiefer, D. Schnoebelen, P. Shafroth, A. H. Siemens, M. C. T. Smith, C. Spence, D. Spencer, M. Taylor, L. Tebay, Tourism Saskatchewan, Travel Montana, U.S. Army Corps of Engineers, Virginia Institute of Marine Science, S. Wade, D. Watkinson, C. White, A. P. Wiens, K. Wilhelm, and G. Winkler.

Both editors wish to express gratitude and appreciation to our wives, Susan Benke and Jackie Cushing, for their encouragement, proofing, and support.

Chapter 1

Introduction

Arthur C. Benke and Colbert E. Cushing

Freshwaters and the rivers that carry them are the continent's most important natural resource in terms of natural biodiversity, a source of water for domestic consumption and irrigation, and various industrial uses. Rivers also happen to be one of the most dramatic features of a continent, are appreciated for their beauty, and often are used for fishing and recreation. They are the inevitable result of precipitation falling across the land, coalescing into streams, and uniting into ever larger streams and rivers. Over millions of years, these networks of flowing waters have delivered sediments and nutrients to downstream areas, sometimes eroding valleys and at other times depositing sediments, before eventually reaching the sea or an inland lake. This movement of water and material has helped shape the terrain, created a diversity of freshwater environments along its path, and allowed the evolution of thousand of species of plants, animals, and microbes. Together, these flowing water environments, with their uniquely adapted species, form the river ecosystems that we see today.

Given the enormous importance of rivers, the basic intention of this book is to present a compact guide to many of the major rivers of North America. This book is based on our *Rivers of North America* (Benke and Cushing 2005), which is a large reference volume of more than 1100 pages. The purpose of our 2005 book was to provide a better understanding of North American rivers and help lead to wiser management, sustainability, and restoration of these essential resources. The purpose of the present book is also to provide a better understanding of rivers, but intended to reach a wider audience. This *Guide* is essentially a distillation of the single-river summaries found in the 2005 book, arranged in the same regional chapters in an easily accessible format.

The North American continent contains a tremendous diversity of river sizes and types. Rivers range from the frigid and often frozen Arctic rivers of northern Canada and Alaska to the warm tropical rivers of southern Mexico. They range from the high-gradient turbulent rivers draining the western mountains to the low-gradient, placid rivers flowing across the southeastern Coastal Plain. River size ranges from what are essentially small streams to the enormous

Mississippi, the 2nd longest river in the world, and the 9th largest by discharge (Leopold 1994). Such variations in latitude, topography, and size contribute to the great variation in biodiversity and ecological characteristics that we see among the continent's rivers.

Total annual discharge from North American rivers is approximately 8,200 km^3/yr or about 17% of the world total (Shiklomanov 1993). The Mississippi is by far the largest river, yet its mean discharge is only 7% of total continental discharge (580 km^3/yr or 18,400 m^3/s) (Shiklomanov 1993, Karr et al. 2000). Among the other top 25 rivers by discharge, more than a dozen have annual discharges greater than 2000 m^3/s, with the largest being the St. Lawrence, Mackenzie, Ohio, Columbia, and Yukon (Table 1). All are rivers that flow to the sea, except the Ohio, which contributes almost half the flow of the Mississippi River. The Nelson and Missouri rivers are among the top five in drainage area, but only rank 11 and 15, respectively, in discharge because their basins receive only moderate precipitation. Three rivers with exceptionally large drainage basins, but not among the top 25 by discharge, are the Colorado, Rio Grande, and Arkansas (see bottom of Table 1). The Colorado and Rio Grande rivers each drain >600,000 km^2 (among the top ten by basin area), but are located in arid regions, and have substantially lower discharge than many rivers draining much smaller basins. In addition to these extremely large rivers and river basins, there are many rivers of moderate-to-large size (100 to >1000 m^3/s) that each flow for several hundred kilometers to the sea or are tributaries of larger rivers.

TABLE 1 Largest rivers of North America ranked by virgin discharge. All rivers may be found in this book except the Koksoak and La Grande.

	River Name	Discharge (m^3/s)	Basin area (km^2)
1	Mississippi	18,400	3,270,000
2	St. Lawrence	12,600	1,600,000
3	Mackenzie	9,020	1,743,058
4	Ohio	8,733	529,000
5	Columbia	7,730	724,025
6	Yukon	6,340	839,200
7	Fraser	3,972	234,000
8	Upper Mississippi	3,576	489,510
9	Slave (Mackenzie basin)	3,437	606,000
10	Usumacinta	2,687	112,550
11	Nelson	2,480	1,072,300

(Continued)

TABLE 1 (Continued)

	River Name	Discharge (m³/s)	Basin area (km²)
12	Liard (Mackenzie basin)	2,446	277,000
13	Koksoak (Quebec)	2,420[1]	133,400[2]
14	Tennessee (Ohio basin)	2,000	105,870
15	Missouri	1,956	1,371,017
16	Ottawa (St. Lawrence basin)	1,948	146,334
17	Mobile	1,914	111,369
18	Kuskokwim	1,900	124,319
19	Churchill (Labrador)	1,861	93,415
20	Copper	1,785	63,196
21	Skeena	1,760	54,400
22	La Grande (Quebec)	1,720[1]	96,866[2]
22	Stikine	1,587	51,592
24	Saguenay (St. Lawrence basin)	1,535	85,500
25	Susitna	1,427	51,800
Additional large basins			
	Rio Grande	~100	870,000
	Colorado	550	642,000
	Arkansas	1,004	414,910

[1]*Dynesius and Nilsson (1994).*
[2]*Leopold (1994).*

Although humans have been attracted to rivers throughout North America for more than 12,000 years, it has not been until the past 100 years that industrialization has caused a radical transformation of most rivers. They have been dammed for flood control, hydropower, and navigation; dewatered for human and agricultural consumption; contaminated with waste products; and invaded by many nonnative species. Such activities have seriously degraded water quality, habitat diversity, biological diversity, and ecosystem integrity of rivers throughout most of the continent. In spite of such extensive alterations, rivers have displayed a remarkable degree of resilience, capable of returning to at least semi-natural conditions when human impacts are reduced. Fortunately, there are still some rivers that have escaped major human alterations, particularly those in the Arctic and Northern Pacific (Chapters 16, 17, and 20). Such pristine or lightly altered rivers retain much of the natural physical and biological properties they have had for millennia, and can serve as benchmarks by which to evaluate impacts and restoration success of altered rivers.

We recognize that modern societies inevitably must exploit rivers for necessary human needs and not all rivers can retain pristine features. However, any objective evaluation of North American rivers would reveal that we have gone well past a balance between human needs and the need for natural riverine ecosystems. Fortunately, the past 40 years have seen a major shift in society's attitudes towards rivers and the need to conserve these valuable natural resources. In spite of progress in our treatment of rivers, however, there have been no efforts in North America to comprehensively evaluate the state of its rivers that is comparable to wetlands evaluations (e.g., see the U.S. Fish and Wildlife Service website for the National Wetlands Inventory). Hopefully, better understanding of North American rivers revealed in this book will help lead to greater appreciation, wiser management, future restoration, and more prolonged sustainability of these essential resources.

INFORMATION IN THIS BOOK

We have selected a total of 200 rivers throughout the continent for this guide, all of which are described in more detail in *Rivers of North America*. Rivers are organized into 22 chapters, some of which are represented by a single major river and its tributaries, such as the Missouri River, and others by region, such as the Atlantic Coast rivers of the Northeastern United States (Fig 1). Material for each chapter was written by regional river experts, and a very condensed version of their text and summary data were retained for this *Guide*. Most of the major rivers and much of the diversity of North American rivers is well represented.

The Introduction to each chapter presents summary text characterizing the river basin or region and includes a map showing the location of the described rivers. This is followed by a series of two-page summaries of the rivers described in that basin or region, each of which contains a color topographic map of the basin showing major tributaries, cities, dams, and boundaries of physiographic provinces. Also included are a color photograph of the river, abbreviated descriptions of physical and biological features, and a graph of mean monthly precipitation/runoff/temperature.

The summary information provided for each river requires some definitions and abbreviated explanations. In some cases, information was not available, or "NA".

Relief

Basin relief is the elevation (in meters) from the highest point in the basin (mountain peak) to the elevation at the river mouth. This statistic, in combination

FIGURE 1   Major river basins and regions used in organization of chapters.

with the total basin area, provides an indication of the general steepness of the river from headwaters to mouth.

Basin area

Basin area is the total surface area (in square kilometers) from which water flows into the river. Along with precipitation and evapotranspiration, basin area determines the total discharge of water at the river's mouth.

Mean annual precipitation

Mean annual precipitation is the mean amount (cm) of precipitation (rain and equivalent of melted snow) averaged over the basin per month (shown in graphs), per year, and averaged over several years (cm per year). It provides an indication of the amount of water provided for any basin, which can vary greatly from arid (<30 cm) to mesic environments (>100 cm).

Mean discharge

Mean discharge is the volume of water flowing out of the mouth of the river in a second of time (cubic meters per second or cubic feet per second), averaged over several years. This is the most common indicator of a river's size. Runoff is discharge divided by basin area and can be expressed as centimeters of water height per month (mean values shown in graphs) or per year. However, natural variation in discharge within any given year can be enormous and vary by more than 100-fold.

Mean air temperature

Mean air temperature is the temperature averaged over the basin and averaged over each month (see graphs) or over several years (degrees Celsius). Air temperature, in combination with precipitation, has an important influence on plant and animal life in the basin and helps determine how much of the precipitation is lost as evapotranspiration rather than flowing into the streams and rivers.

Mean water temperature

Mean water temperature is the temperature averaged over the year. It typically is similar to the mean air temperature, but is less subject to extreme high and low values throughout the year. Water temperature plays an important role in determining what species can exist in any river.

Physiographic provinces

Physiographic provinces are broad scale subdivisions of the continent based on topographic features, rock type, and geological structure and history. See Figure 2 for a map of physiographic provinces of the entire continent. Among the larger provinces are Great Plains, Coastal Plain, and Central Lowland. Physiographic provinces for each river basin are shown in the river summary, along with corresponding acronyms; e.g., Great Plains (GP), Coastal Plain (CP), etc. These acronyms are used to identify each province on the color topographic maps, and if there are two or more provinces, they are separated by yellow lines.

Number of fish species

Fish diversity is the best known of any aquatic group, with more than 1000 species of native freshwater fishes throughout North America. For many rivers, we have a good idea of the number of species present. This diversity varies greatly across the continent, with a particularly high concentration of species occurring in the southeastern United States, and often over 100 species per river.

1. Arctic Slope
2. Brooke Range
3. Mackenzie Mts.
4. Seward Pen. and
   Bering Coast Uplands
5. Yukon Basin
6. Alaskan Pen. and
   Aleutian Islands
7. South Central Alaska
8. Arctic Lowlands
9. Baffin Upland
10. Coast Mountains of BC
    and SE Alaska
11. Rocky Mountains in
    Canada
12. Great Plains
13. Thelon Plains and Bear
    River Lowland
14. Bear-Slave-Churchill
    Uplands
15. Athabasca Plain
16. Hudson Bay Lowland
17. Labrador Highlands
18. Superior Upland
19. Laurentian Highlands
20. St. Lawrence Lowland
21. New England/Maritime

22. Adirondack
23. Coastal Plain
24. Piedmont Plateau
25. Blue Ridge
26. Valley and Ridge
27. Appalachian Plateaus
28. Interior Low Plateaus
29. Central Lowland
30. Ozark Plateaus
31. Ouachita

32. Southern Rocky Mts.
33. Wyoming Basin
34. Colorado Plateaus
35. Middle Rocky Mts.
36. Northern Rocky Mts.
37. Columbia Plateau

38. Basin and Range
39. Cascade-Sierra Mts.
40. Pacific Border
41. Lower California
42. Baja California
43. Buried Ranges
44. Sierra Madre Occidental
45. Sierra Madre Oriental
46. Neovolcanic Plateau
47. Sierra Madre Del Sur
    System
48. Chiapas-Guatemala
    Highlands
49. Yucatan

FIGURE 2   Physiographic provinces of North America.

Number of threatened or endangered species

A large number of species are at risk of extinction, often being designated either threatened or endangered. This is particularly true in the southeastern and southwestern U.S., and in Mexico, and most of these threatened or endangered species are either fishes or mollusks. Knowledge of which rivers contain endangered species heightens the importance of reducing pollution, maintaining natural flow regimes, and retaining natural habitats in those rivers.

Major fishes

Rivers are often characterized by the composition and relative abundance of their fish species, which are commonly used as indicators of water quality. This fish composition varies greatly across the continent. Because there may be >100 species in some rivers, we list only the most common.

Major other aquatic vertebrates

Besides fishes, rivers are home to many species of amphibians, reptiles, birds, and mammals. These species are not always as well known as the fishes, and may be transitory, but they can be an important part of food webs. We list only a few of the most common species.

Major benthic insects

Benthic invertebrates (particularly aquatic insects, mollusks, and crustaceans) live on the bottom (benthic habitat) of rivers and are commonly used as indicators of water quality. There are typically more species of invertebrates than fishes, but the total number is rarely assessed. We only list some of the more common aquatic insects among the mayflies, caddisflies, stoneflies, and true flies.

Nonnative species

Nonnative species are typically aquatic, particularly fishes, that have been introduced from outside the river basin. When abundant, they can greatly reduce or eliminate native species, and are thus considered undesirable by conservationists. We list the more problematic nonnative species found in the river.

Major riparian plants

Plants, particularly tree species, grow along the river's bank and in its floodplain, and are very dependent on the river's natural flooding patterns. They stabilize bank and floodplain environments and are an important part of the habitat and food supply for aquatic animals.

Special features

We list a few of the characteristics that distinguish each river such as major wetlands (floodplains), waterfalls, long free-flowing reaches, protected areas, and a particularly high diversity of species. Such features enhance their conservation value as well as help preserve their natural biodiversity and ecosystem function.

Fragmentation

Fragmentation represents the degree to which a river has been subdivided by dams built for power generation, water supply, flood control, and recreation.

Dams typically result in a dramatic alteration of natural discharge patterns and sediment loads that considerably degrade the river's natural features. Dams can sometimes reduce a river's flow to zero with devastating ecological consequences.

CONCLUDING COMMENTS

Readers of this *Guide* should realize that the river summaries presented here are only a brief snapshot of each river. Even though additional information is presented in *Rivers of North America* (e.g., water quality and land use information), it should be recognized that in spite of all the information presented in these volumes, our understanding of natural biodiversity and ecology in North American rivers is fair at best. Biological, ecological, and water quality data for many rivers are superficial, and sometimes non-existent. It is difficult to assess the natural biological characteristics of most rivers, now that they are already highly modified or polluted. On the other hand, most of the remaining pristine rivers are at such remote locations (Arctic) that little research has been conducted on them. Clearly, there are large gaps in our biological knowledge. There is a great need for major new research initiatives that can lead to better scientific understanding, wider appreciation of their importance, and wiser management. In spite of this large gap in scientific knowledge of rivers, it is imperative that our current knowledge be put to good use in conservation and management if the natural features and biodiversity of rivers are to be retained.

LITERATURE CITED

Benke, A. C., & Cushing, C. E. (Eds.). (2005). *Rivers of North America*. Burlington, MA: Academic Press/Elsevier.

Dynesius, M., & Nilsson, C. (1994). Fragmentation and flow regulation of river systems in the northern third of the world. *Science, 266*, 753–762.

Karr, J. R., Allan, J. D., & Benke, A. C. (2000). River conservation in the United States and Canada. In P. J. Boon, B. R. Davies, & G. E. Petts (Eds.), *Global perspectives on river conservation: science, policy and practice* (pp. 3–39). Chichester, England: John Wiley & Sons.

Leopold, L. B. (1994). A view of the river. Cambridge, Massachusetts: Harvard University Press.

Shiklomanov, I. A. (1993). Chapter 2. World fresh water resources. In P. H. Gleick (Ed.), *Water in crisis: a guide to the world's fresh water resources* (pp. 13–24). Oxford University.

Atlantic Coast Rivers of the Northeastern United States

John K. Jackson, Alexander D. Huryn, David L. Strayer, David L. Courtemanch, and Bernard W. Sweeney

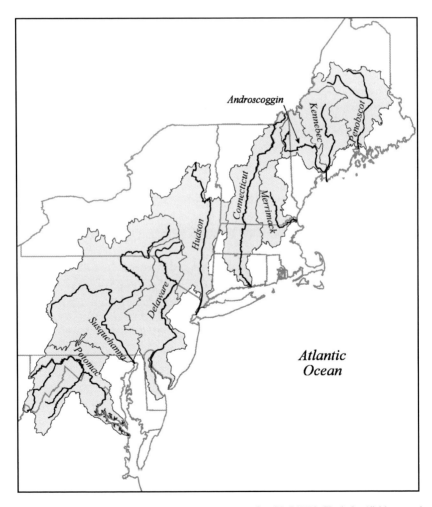

The Atlantic slope region of the northeastern United States stretches from the Penobscot River in northern Maine (46°N) to the Rappahannock River on the lower Chesapeake Bay (37.5°N). The larger rivers in this region flow south or southeast between the Appalachian Mountains and the Atlantic Ocean (see map). The Appalachian Mountains form the western border of the region and include such well-known ranges as the White Mountains in New Hampshire, the Green Mountains in Vermont, the Catskill Mountains in New York, and the Allegheny Mountains in Pennsylvania and West Virginia. Elevations in the mountains and uplands are frequently 600 to 800 m asl, with peaks reaching 1100 to >1600 m asl. The relatively short distance between the mountains (or other basin divides) and the coast limits both basin area and river length in the region, while the region's abundant precipitation and high runoff result in high average discharge relative to basin area.

The river basins of the northeast have long histories of human occupation, with archaeological evidence at some locations dating back at least 11,000 years. The rivers played important roles in the European colonization of North America and the establishment of the United States. The basins were dominant features in 10 of the original 13 English colonies, and they contributed vast quantities of lumber, minerals, fish, agricultural products, and power for both local consumption and export. The rivers themselves served as economic, transportation, and communication conduits, and early settlers established such cities as Boston, New York, Philadelphia, and Baltimore on sites where there were natural ports and river access. While urban, suburban, and industrial development have had major impacts on the streams and rivers in the northeast, agriculture, silviculture, mining, dams and diversions, and nonnative species can also significantly affect the quality and quantity of these fresh water systems.

There are six large river basins ($>20,000 \text{ km}^2$) in the Atlantic U.S. Northeast region with mean discharge $>300 \text{ m}^3/\text{s}$, all of which are described in this chapter: Penobscot, Connecticut, Hudson, Delaware, Susquehanna, and Potomac. The largest river in the region is the Susquehanna (basin area of $>71,000 \text{ km}^2$ and mean discharge $>1100 \text{ m}^3/\text{s}$), which is the 22nd largest river in North America, and the 3rd largest in North America that flows into the Atlantic Ocean [after the St. Lawrence and Churchill (Labrador)]. The Hudson River is one of the most intensively studied rivers in North America, making it one of the best examples to illustrate the physical, chemical, and biological complexity characteristic of large river basins. Features of several smaller river basins ($<15,000 \text{ km}^2$) are also described (Kennebec, Androscoggin, Merrimack).

Penobscot River

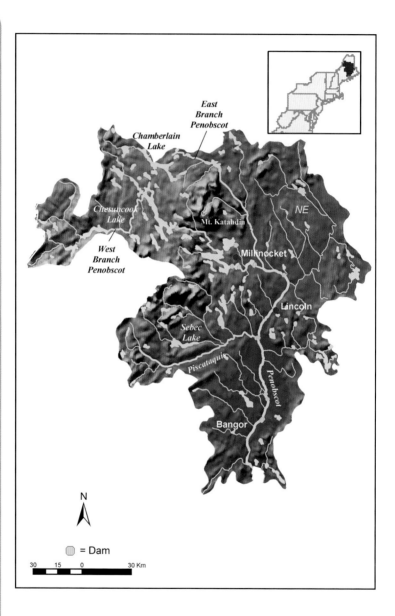

East
Branch
Penobscot

Chamberlain
Lake

Chesuncook
Lake

Mt. Katahdin

NE

Millinocket

West
Branch
Penobscot

Lincoln

Sebec
Lake

Piscataquis

Penobscot

Bangor

N

⬤ = Dam

30   15   0   30 Km

Relief: 1607 m
Basin area: 22,253 km^2
Mean discharge: 402 m^3/s
Mean annual precipitation: 107 cm

Mean air temperature: 9.5°C
Mean water temperature: 9.3°C
No. of fish species: 45
No. of endangered species: 1

Physiographic province: New England (NE)

Major fishes: alewife, American eel, American shad, Atlantic salmon, brown bullhead, burbot, chain pickerel, common shiner, creek chub, fallfish, pumpkinseed, redbreast sunfish, smallmouth bass, white sucker, white perch, yellow perch

Major other aquatic vertebrates: eastern painted turtle, snapping turtle, muskrat, beaver, mink, river otter

Major benthic insects: mayflies (*Centroptilum, Ephemerella, Eurylophella, Heptagenia, Stenonema*), stoneflies (*Acroneuria*), caddisflies (*Cheumatopsyche, Hydropsyche, Macrostemum, Neureclipsis, Polycentropus*)

Nonnative species: chain pickerel, smallmouth bass, brown trout

Major riparian plants: balsam fir, red maple, silver maple, paper birch, black ash, quaking aspen,

chokecherry, northern white cedar, American elm (declining)

Special features: River islands north of Old Town comprise Penobscot Indian reservation; Baxter State Park and Mt. Katahdin; Sunkhaze National Wildlife Refuge

Fragmentation: 5 major dams span main stem; 111 additional licensed dams (many hydroelectric) on tributaries.

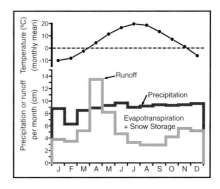

Penobscot River at Five Islands Rapids, below Mattawamkeag, Maine (photo by Tim Palmer).

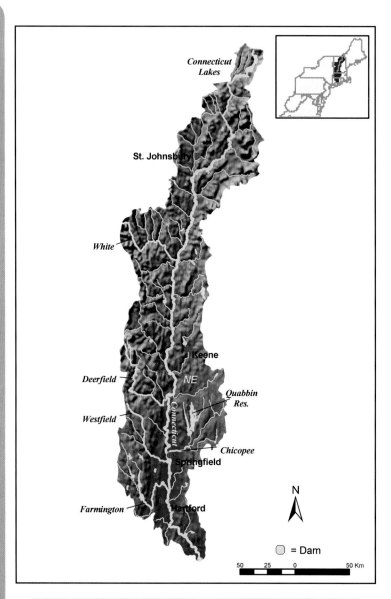

Connecticut River

Relief: 1917 m
Basin area: 29,160 km²
Mean discharge: 445 m³/s
Mean annual precipitation: 109 cm

Mean air temperature: 10.5°C
Mean water temperature: NA
No. of fish species: 64 fresh-
water, 44 estuarine
No. of endangered species: 2

Physiographic province: New England (NE)

Major fishes: American shad, blueback herring, sea lamprey, striped bass, brook trout, American eel, white sucker, yellow perch, fallfish, common shiner, golden shiner, spottail shiner, banded killifish, redbreast sunfish, pumpkinseed, brown bullhead

Other aquatic vertebrates: beaver, river otter, northern water snake, snapping turtle, bald eagle, bank swallow, common loon, common merganser, belted kingfisher, great blue heron

Major benthic insects: stoneflies (*Acroneuria*), mayflies (*Eurylophella, Serratella, Stenonema*), caddisflies (*Brachycentrus, Chimarra, Hydropsyche, Neureclipsis, Oecetis*), true flies (*Chironomus, Polypedilum, Microtendipes, Glyptotendipes, Tanytarsus*)

Nonnative species: 20 fish species (brown trout, smallmouth bass, large-mouth bass, black crappie, white crappie, northern pike, common carp, rainbow trout, bowfin, bluegill), Asiatic clam

Major riparian plants: cord grass, wild celery, broadleaf arrowhead, cattail, pickerelweed, purple loosestrife

Special features: Watershed, including tidal wetlands, has received national and international recognition of its ecological uniqueness and value.

Fragmentation: 16 hydroelectric dams

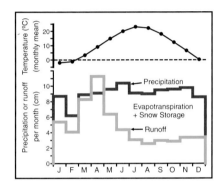

Connecticut River near Bloomfield, Vermont (photo by Tim Palmer).

Hudson River

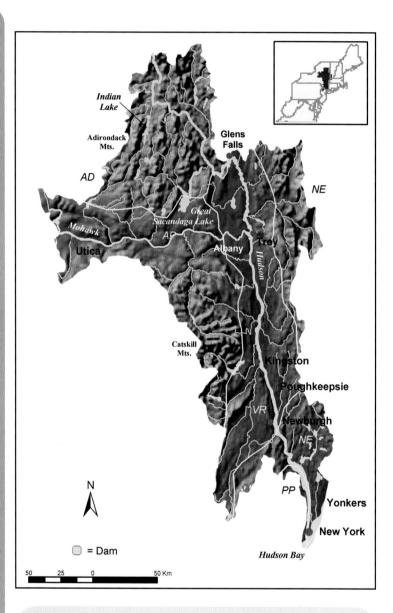

Indian Lake
Adirondack Mts.
Glens Falls
AD
NE
Great Sacandaga Lake
Mohawk
AP
Utica
Albany
Troy
Hudson
Catskill Mts.
Kingston
Poughkeepsie
VR
Newburgh
NE
N
PP
Yonkers
New York
⬤ = Dam
Hudson Bay

50 25 0   50 Km

Relief: 1629 m
Basin area: 34,615 km²
Mean discharge: 592 m³/s
Mean annu precipitation: 92 cm

Mean air temp.: 8.9°C
Mean water temp.: 12.4°C
No. of fish species: >200 (70 freshwater, 95 brackish)
No. of endangered species: 1

Physiographic provinces: Adirondack Mountains (AD), Valley and Ridge (VR), Appalachian Plateau (AP), New England (NE), Piedmont Plateau (PP)
Major fishes: American eel, American shad, blueback herring, alewife, white catfish, white sucker, common carp, cutlips minnow, blacknose dace, creek chub, fallfish, common shiner, spottail shiner, brook trout, Atlantic tomcod, banded killifish, striped bass, white perch, smallmouth bass, largemouth bass, tessellated darter
Major other aquatic vertebrates: snapping turtle, mute swan, Canada goose, beaver
Major benthic insects: mayflies (*Stenonema, Caenis*), caddisflies (*Neureclipsis, Chimarra, Hydropsyche, Cheumatopsyche*)
Nonnative species: curly-leaved pondweed, water-chestnut, Eurasian watermilfoil, purple loosestrife, mud bithynia, zebra mussel, dark

falsemussel, Atlantic rangia, common carp, brown trout, northern pike, rock bass, smallmouth bass, largemouth bass, black crappie, mute swan
Major riparian plants (freshwater tidal Hudson): silver maple, red maple, cottonwood, sycamore, willows, common reed, narrowleaf cattail
Special features: Drains much of Adirondack Mountains; long (248 km) intertidal zone; anadromous fishery
Fragmentation: 14 dams in middle section; no dams in upper river

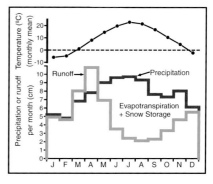

Hudson River above West Point, New York (photo by Tim Palmer).

Delaware River

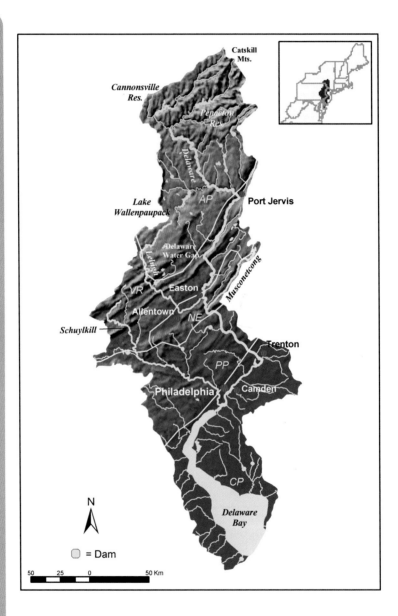

N

⬤ = Dam

| 50 | 25 | 0 | 50 Km |

Relief: 698 m
Basin area: 33,041 km^2
Mean discharge: 422 m^3/s
Mean annual precipitation: 108 cm

Mean air temperature: 11.8°C
Mean water temperature: 13.8°C
No. of fish species: 105
No. of endangered species: 5

Physiographic provinces: Appalachian Plateau (AP), Valley and Ridge (VR), New England (NE), Piedmont Plateau (PP), Coastal Plain (CP)

Major fishes: American eel, American shad, black crappie, blue catfish, brown bullhead, carp, fallfish, river chub, river herring, rainbow trout, redbreast sunfish, rock bass, striped bass, white perch, yellow perch

Major other aquatic vertebrates: northern water snake, snapping turtle, eastern mud turtle, painted turtle, spotted turtle, red bellied turtle, beaver, river otter

Major benthic insects: caddisflies (*Brachycentrus*, *Cheumatopsyche*, *Chimarra*, *Hydropsyche*, *Lepidostoma*, *Rhyacophila*), mayflies (*Acentrella*, *Baetis*, *Epeorus*, *Eurylophella*, *Isonychia*, *Stenonema*, *Tricorythodes*), stoneflies (*Acroneuria*)

Nonnative species: Brazilian waterweed, brittle naiad, Carolina fanwort, dotted duckweed, Eurasian watermilfoil, European water clover, hydrilla, parrot feather, pond water starwort, purple loosestrife, sacred lotus, watercress, water lettuce, Asiatic clam, bluegill, brown trout, carp, channel catfish, largemouth bass, smallmouth bass, rock bass, walleye

Major riparian tree species: American beech, American hornbeam, bitternut hickory, sweet birch, black cherry, black locust, black walnut, butternut, eastern hemlock, catalpa, chestnut oak, hackberry, pignut hickory, red elm, red maple, northern red oak, silver maple, sour gum, sugar maple, tulip poplar, white ash, white oak

Special features: Longest undammed main-stem river (530 km) in eastern U.S.; several km of main stem designated National Wild and Scenic River; Delaware Water Gap

Fragmentation: 16 major dams on tributaries

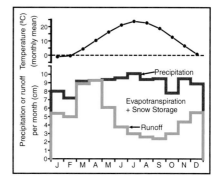

Delaware River above Belvidere, New Jersey (photo by A. C. Benke).

Susquehanna River

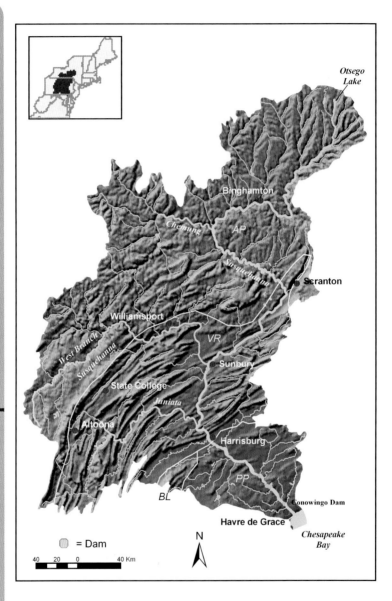

Relief: 959 m
Basin area: 71,432 km²
Mean discharge: 1153 m³/s
Mean annual precipitation: 98 cm

Mean air temperature: 9.7°C
Mean water temperature: 14°C
No. of fish species: 103
No. of endangered species: 2

Physiographic provinces: Appalachian Plateau (AP), Valley and Ridge (VR), Piedmont Plateau (PP), Blue Ridge (BL)

Major fishes: American eel, American shad, blueback herring, alewife, rock bass, smallmouth bass, channel catfish, walleye, muskellunge, striped bass, spotfin shiner, banded darter, bluntnose minnow, margined madtom, bluegill

Major other aquatic vertebrates: beaver, muskrat, common map turtle, eastern painted turtle, snapping turtle, wood turtle, northern water snake, great blue heron, bald eagle, peregrine falcon, mallard duck, wood duck, Canada goose

Major benthic insects: mayflies (*Baetis*, *Pseudocloeon*, *Centroptilum*, *Stenonema*, *Leucrocuta*, *Isonychia*, *Serratella*, *Anthopotamus*), caddisflies (*Hydropsyche*, *Cheumatopsyche*, *Hydroptila*), stoneflies (*Agnetina*), true flies (*Rheotanytarsus*, *Polypedilum*, *Tvetenia*, *Chironomus*, *Dicrotendipes*)

Nonnative species: 27 fishes (smallmouth bass, channel catfish), Asiatic clam, zebra mussel, rainbow mussel, rusty crayfish, watercress, Eurasian watermilfoil, purple loosestrife, curly pondweed, Japanese knotweed, mile-a-minute weed

Major riparian plants: sycamore, tulip poplar, red maple, silver maple, river birch, black willow, black cherry, American beech, black locust

Special features: Broad river channel, numerous islands, largest commercially non-navigable river in U.S.

Fragmentation: >100 dams, first major dam 16 km from mouth

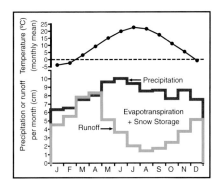

Susquehanna River above Harrisburg, Pennsylvania (photo by Tim Palmer).

Kennebec River

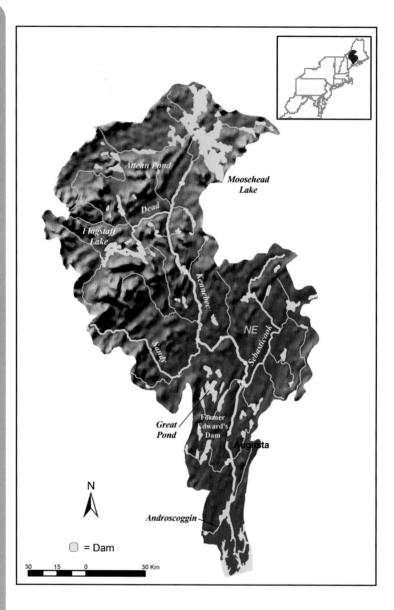

Relief: 1234 m
Basin area: 13,944 km²
Mean discharge: 257 m³/s
Mean annual precipitation: 108 cm

Mean air temperature: 11.9°C
Mean water temperature: 10.2°C
No. of fish species: 48
No. of endangered species: 1

Physiographic province: New England (NE)

Major fishes: alewife, blueback herring, Atlantic salmon (including migratory and landlocked populations), American shad, striped bass, rainbow smelt, smallmouth bass, brook trout, brown trout, rainbow trout, white perch, yellow perch, white sucker, fallfish

Major other aquatic vertebrates: eastern painted turtle, snapping turtle, beaver, muskrat, river otter

Major benthic insects: mayflies (*Ameletus, Caenis, Epeorus, Ephemerella, Eurylophella, Heptagenia, Stenacron, Stenonema*), caddisflies (*Cheumatopsyche, Chimarra, Macrostemum, Mystacides, Neureclipsis, Polycentropus*)

Nonnative species: purple loosestrife, rusty crayfish, common carp, white catfish, gizzard shad, brown trout, rainbow trout, smallmouth bass, largemouth bass, black crappie, northern pike, European rudd

Major riparian plants: balsam fir, red maple, paper birch, eastern white pine, chokecherry, northern red oak, eastern hemlock

Special features: First major dam (Edwards Dam) on major river breached by U.S. government (1999); Moosehead Lake ($311 km^2$); Merrymeeting Bay—large freshwater tidal estuary

Fragmentation: 8 hydroelectric dams

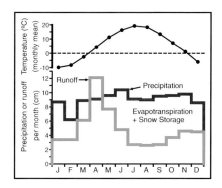

Kennebec River at Solon, Maine (photo by Tim Palmer).

Androscoggin River

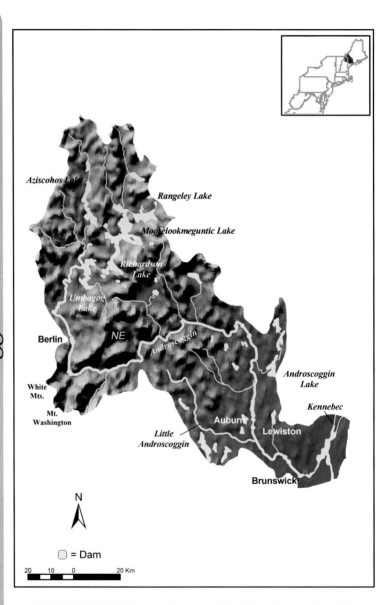

Aziscohos Lake

Rangeley Lake

Mooselookmeguntic Lake

Richardson Lake

Umbagog Lake

Berlin   *NE*

Androscoggin

White Mts.

Androscoggin Lake

Mt. Washington

Kennebec

Auburn

Lewiston

Little Androscoggin

Brunswick

N

◯ = Dam

20 10 0   20 Km

Relief: 1234 m
Basin area: 8451 km²
Mean discharge: 175 m³/s
Mean annual precipitation: 111 cm

Mean air temperature: 10.9°C
Mean water temperature: 9.7°C
No. of fish species: 33
(27 native), 7 estuarine
No. of endangered species: 1

Physiographic province: New England (NE)

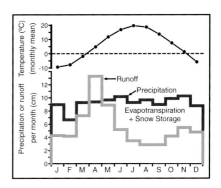

Major fishes: brown trout, rainbow trout, smallmouth bass, white perch, yellow perch, white sucker, fallfish

Major other aquatic vertebrates: beaver, muskrat, river otter

Major benthic insects: mayflies (*Ameletus*, *Caenis*, *Ephemerella*, *Eurylophella*, *Heptagenia*, *Stenacron*, *Stenonema*), caddisflies (*Cheumatopsyche*, *Chimarra*, *Macrostemum*, *Mystacides*, *Neureclipsis*, *Polycentropus*)

Nonnative species: purple loosestrife, rusty crayfish, calico crayfish, brown trout, rainbow trout, smallmouth bass, largemouth bass, northern pike, spottail shiner, common carp, white catfish

Major riparian plants: balsam fir, red maple, paper birch, American witch hazel, eastern white pine, chokecherry, northern red oak, eastern hemlock

Special features: Lake Umbagog National Wildlife Refuge

Fragmentation: 14 hydroelectric dams from source at Umbagog Lake to tidewater

Androscoggin River near Wilsons Mills, Maine (photo by Tim Palmer).

Merrimack River

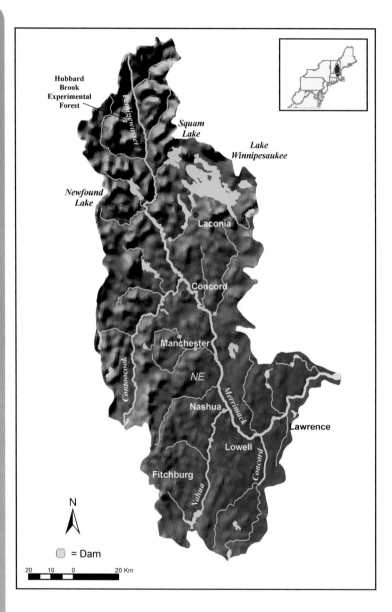

Relief: 1563 m
Basin area: 12,986 km²
Mean discharge: 235 m³/s
Mean annual precipitation: 92 cm

Mean air temperature: 7.3°C
Mean water temperature: NA
No. of fish species: 50
No. of endangered species: 1

Physiographic province: New England (NE)

Major fishes: American eel, alewife, brook trout, brown trout, rainbow trout, chain pickerel, fallfish, common shiner, white sucker, white perch, smallmouth bass, largemouth bass, black crappie, bluegill

Major benthic insects: caddisflies (Hydropsychidae, Philopotamidae, Leptoceridae), true flies (chironomid midges)

Nonnative species: common carp, brown trout, smallmouth bass, large-mouth bass, black crappie

Major riparian plants: NA

Fragmentation: high (>500 dams in basin, several on main stem)

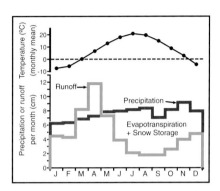

Merrimack River near Franklin, New Hampshire (photo by Tim Palmer).

Potomac River

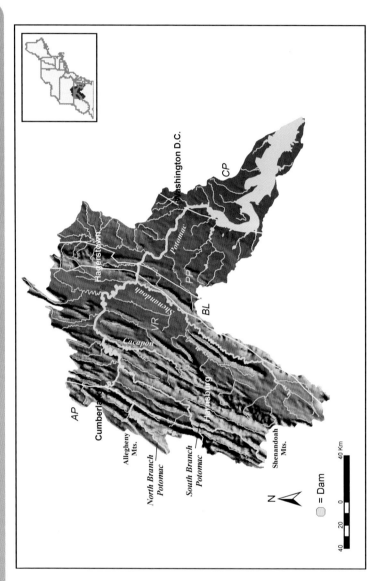

Relief: 1481 m
Basin area: 37,995 km²
Mean discharge: 320 m³/s
Mean annual precipitation: 99 cm

Mean air temperature: 12.3°C
Mean water temperature: 14.0°C
No. of fish species: 95 species
(65 native)
No. of endangered species: 2

Physiographic provinces: Appalachian Plateau (AP), Valley and Ridge (VR), Piedmont Plateau (PP), Blue Ridge (BL), Coastal Plain (CP)

Major fishes: smallmouth bass, channel catfish, spottail shiner, spotfin shiner, bluntnose minnow, redhorse sucker, redbreast sunfish, bluegill, tessellated darter

Major other aquatic vertebrates: beaver, muskrat, common snapping turtle, eastern painted turtle, great blue heron, bald eagle, osprey, Canada goose

Major benthic insects: mayflies (*Anthopotamus, Caenis, Ephoron, Serratella, Stenonema, Tricorythodes*), caddisflies (*Hydropsyche, Cheumatopsyche, Hydroptila, Potamyia, Macrostemum*), true flies (*Thienemannimyia, Tanytarsus, Cricotopus, Polypedilum, Cryptochironomus, Synorthocladius*)

Nonnative species: smallmouth bass, channel catfish, common carp, northern pike, muskellunge, Asiatic clam, hydrilla, Eurasian watermilfoil

Special features: second-largest tributary to Chesapeake Bay, 174 km tidal/estuarine section

Fragmentation: High, but only 3 impoundments >4 km²; >60 blockages in the Anacostia basin being removed or altered for fish passage.

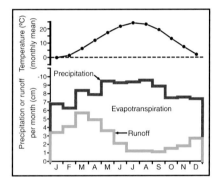

Potomac River near Harper's Ferry, West Virginia (photo by A. C. Benke).

Chapter 3

Atlantic Coast Rivers of the Southeastern United States

Leonard A. Smock, Anne B. Wright, and Arthur C. Benke

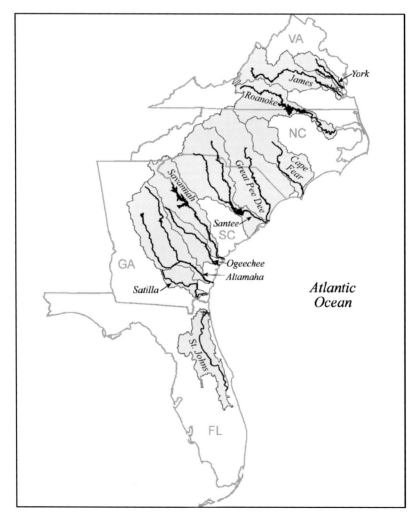

DOI: 10.1016/B978-0-12-375088-4.00003-8

The Atlantic slope region of the southeastern United States encompasses a broad geographic area from 38°N to 26°N latitude, ranging from central Virginia to eastern Florida (see map). The northern part of the region includes the Southern Appalachian Mountains along its western boundary to the flat coastal areas along the Atlantic Ocean. Below 31°N latitude in south Georgia and Florida, however, there is only Coastal Plain. The region has abundant rainfall, moderate to warm temperatures, historically dense forests, and a landscape mosaic of forests, wetlands, agriculture, and urbanized areas. A series of rivers drain the region, most flowing in a primarily southeasterly direction to the Atlantic coast. The diverse geology, physiography, and climate of the region have produced rivers that have variable geomorphic, hydrologic, and biological characteristics. Many of the larger rivers have their headwaters in the rugged Appalachian Mountains, whereas others originate among rolling Piedmont hills or on the flat Coastal Plain.

Flowing here are some of the most historic rivers on the continent. They had a profound influence on Native American civilizations that occupied these basins for over 11,000 years until well into the eighteenth century. The first Europeans to occupy this region were the Spanish missionaries who settled along the coast in the early 1500s and remained for more than a century. The establishment of Jamestown in 1607, along the James River in Virginia, marked the British arrival in North America that would eventually push Native Americans out of the southeastern Atlantic region. British colonists established towns at the mouths of major rivers that would eventually become major cities. They used the rivers as corridors for inland incursions that would help develop the agricultural and commercial strength of the United States.

Sixteen major rivers are found along the southeastern Atlantic slope, from the York River on the northernmost border to rivers on the eastern coast of Florida. The mouths of major rivers occur about every 100 km along the coast (see map). This chapter describes 11 of these rivers, including the largest with drainage areas of $>20,000 \text{ km}^2$: James, Roanoke, Cape Fear, Great Pee Dee, Santee, Savannah, Altamaha, and St. Johns. The largest river is the Santee, with a basin area of almost $40,000 \text{ km}^2$ and mean discharge $>400 \text{ m}^3/\text{s}$. Other smaller rivers covered in this chapter include the York, Ogeechee, and Satilla. Rivers of the southeastern Atlantic have lower discharge for a given basin area (i.e., lower runoff) than Atlantic rivers of the northeastern U.S. and Canada due to much higher evapotranspiration in a warmer climate.

James River

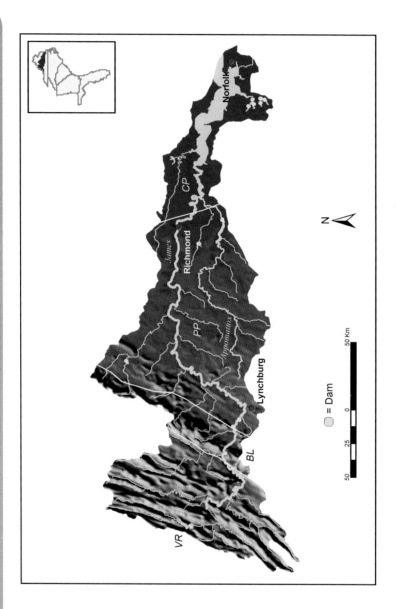

Norfolk

CP

James

Richmond

PP

Appomattox

Lynchburg

BL

VR

N

50 Km

= Dam

50   25   0   50

Relief: 1250 m

Basin area: 26,164 km^2

Mean discharge: 227 m^3/s

Mean annual precipitation: 108 cm

Mean air temperature: 14°C

Mean water temperature: 16°C

No. of fish species: 109

No. of endangered species: 11

Physiographic provinces: Valley and Ridge (VR), Blue Ridge (BL), Piedmont Plateau (PP), Coastal Plain (CP)

Major fishes: American eel, American shad, hickory shad, blueback herring, alewife, gizzard shad, common carp, bull chub, quillback sucker, satinfin shiner, spottail shiner, flathead catfish, blue catfish, white perch, striped bass, redbreast sunfish, bluegill, smallmouth bass, largemouth bass

Major other aquatic vertebrates: cottonmouth, water snakes, painted turtle, musk turtle, river cooter, red-bellied turtle, snapping turtle, muskrat, beaver

Major benthic insects: mayflies (*Tricorythodes*, *Stenonema*, *Baetis*, *Caenis*), caddisflies (*Hydropsyche*, *Cheumatopsyche*, *Polycentropus*), true flies (*Rheotanytarsus*)

Nonnative species: Asiatic clam, smallmouth bass, largemouth bass, rock bass, bluegill, flathead catfish, blue catfish, channel catfish, muskellunge, walleye, threadfin shad

Major riparian plants: sycamore, swamp black gum, river birch, American elm, red maple, ash-leaf maple, bald cypress

Special features: drains from 4 physiographic provinces, no large impoundments, class IV white-water rapids at Fall Line

Fragmentation: 12 low-head dams, some partially breached, that regulate flow in about 10% of nontidal river

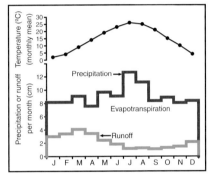

James River at Blue Ridge Parkway, northwest of Lynchburg, Virginia (photo by Tim Palmer).

Cape Fear River

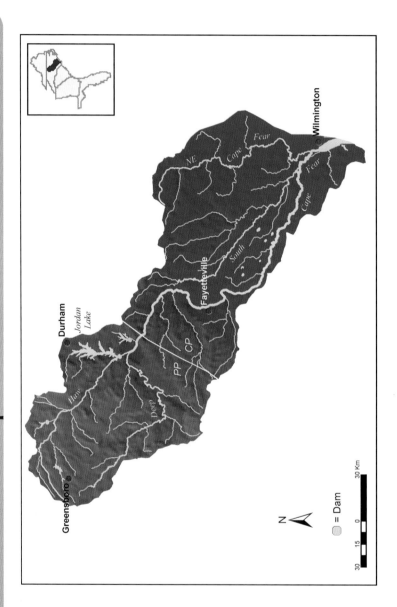

Relief: 305 m
Basin area: 24,150 km²
Mean discharge: 217 m³/s
Mean annual precipitation: 119 cm

Mean air temperature: 16°C
Mean water temperature: 17°C
No. of fish species: 95
No. of endangered species: 24

Physiographic provinces: Piedmont Plateau (PP), Coastal Plain (CP)

Major fishes: American eel, American shad, hickory shad, blueback herring, gizzard shad, common carp, spotted sucker, shiners, darters, channel catfish, bluegill, pumpkinseed, largemouth bass, striped bass

Major other aquatic vertebrates: cottonmouth, water snakes, painted turtle, musk turtle, river cooter, slider, mud turtle, snapping turtle, muskrat, bull frog, river otter, beaver

Major benthic insects: mayflies (*Tricorythodes, Caenis, Stenonema, Baetis*), caddisflies (*Hydropsyche, Cheumatopsyche*)

Nonnative species: Asiatic clam, smallmouth bass, white crappie, flathead catfish, channel catfish

Major riparian plants: sycamore, sweetgum, swamp black gum, red maple, bald cypress, water tupelo, ashes, oaks

Special features: lower tributaries (Black River, Northeast Cape Fear River) are classic southeastern blackwater rivers with broad hardwood floodplains; Carolina Bay lakes in basin

Fragmentation: 3 locks and dams on main stem; large impoundment on primary tributary (Haw River)

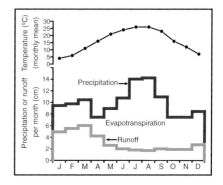

Cape Fear River near Erwin, North Carolina (photo by Tim Palmer).

Savannah River

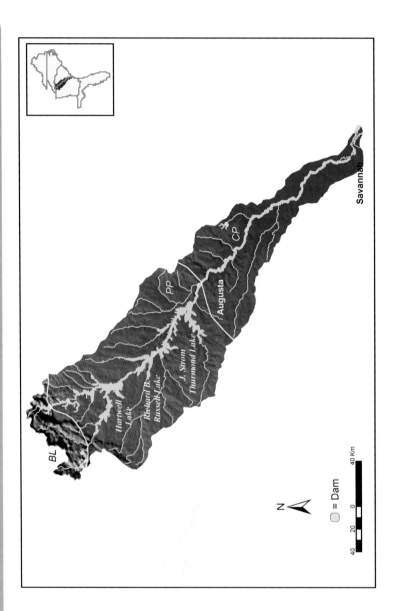

Relief: 1743 m
Basin area: 27,414 km^2
Mean discharge: 319 m^3/s
Mean annual precipitation: 114 cm

Mean air temperature: 18°C
Mean water temperature: 17°C
No. of fish species: 106
No. of endangered species: 24

Physiographic provinces: Blue Ridge (BL), Piedmont Plateau (PP), Coastal Plain (CP)

Major fishes: bowfin, American eel, American shad, blueback herring, common carp, eastern silvery minnow, shiners, silver redhorse, spotted sucker, channel catfish, striped bass, black crappie, redbreast sunfish, bluegill, largemouth bass

Major other aquatic vertebrates: American alligator, cottonmouth, water snakes, snapping turtle, musk turtle, river cooter, slider, river frog, muskrat, river otter, beaver

Major benthic insects: mayflies (*Baetis*, *Caenis*, *Isonychia*, *Stenonema*, *Tricorythodes*), caddisflies (*Cheumatopsyche*, *Chimarra*, *Hydropsyche*, *Macrostemum*, *Neureclipsis*, *Oecetis*), true flies (*Rheotanytarsus*, *Simulium*)

Nonnative species: hydrilla, waterweed, Asiatic clam, flathead catfish, channel catfish

Major riparian plants: bald cypress, water tupelo, swamp black gum, sweetgum, water hickory, red maple, sycamore, oaks

Special features: broad forested floodplain swamp throughout the Coastal Plain

Fragmentation: several large dams in Piedmont

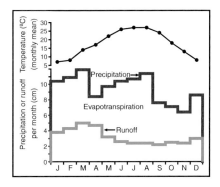

Savannah River below Thurmond Dam, upstream of Augusta, Georgia (photo by A. C. Benke).

Ogeechee River

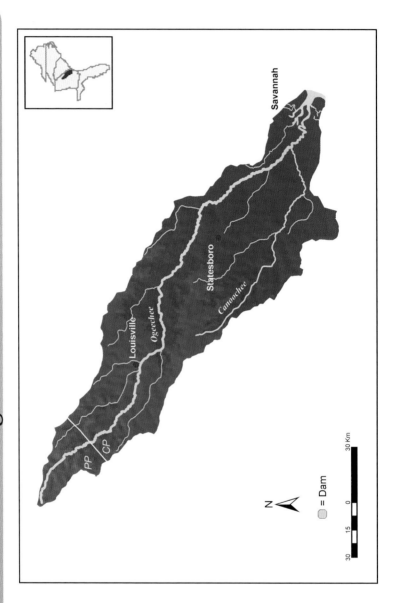

Relief: 200 m   Mean air temperature: 18°C
Basin area: 13,500 km²   Mean water temperature: 19°C
Mean discharge: 115 m³/s   No. of fish species: >80
Mean annual precipitation: 113 cm   No. of endangered species: 11

Physiographic provinces: Piedmont Plateau (PP), Coastal Plain (CP)
Major fishes: American eel, longnose gar, bowfin, snail bullhead, redbreast sunfish, spotted sunfish, bluegill, largemouth bass, chain pickerel, spotted sucker, shiners, American shad, black crappie, warmouth, redear sunfish, white catfish, chubs, darters, silversides
Major other aquatic vertebrates: cottonmouth, water snakes, softshell turtles, river cooter, American alligator, river frog, sirens, treefrogs, dusky salamanders, muskrat, river otter, beaver
Major benthic insects: caddisflies (*Hydropsyche, Cheumatopsyche, Hydroptila, Chimarra*), stoneflies (*Perlesta, Paragnetina, Taeniopteryx, Pteronarcys*), mayflies (*Stenonema, Baetis, Isonychia, Ephemerella, Tricorythodes, Caenis*), true flies (*Polypedilum, Rheotanytarsus*)
Nonnative species: Asiatic clam

Major riparian plants: swamp black gum, water tupelo, bald cypress, red maple, water oak, laurel oak, sweetgum, water hickory
Special features: broad forested floodplain swamp ($>$1 km width); extensive submerged wood habitat; one of few natural free-flowing rivers in coterminous 48 states
Fragmentation: no major dams

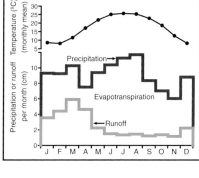

Ogeechee River, upstream of I-16 (photo by A. C. Benke).

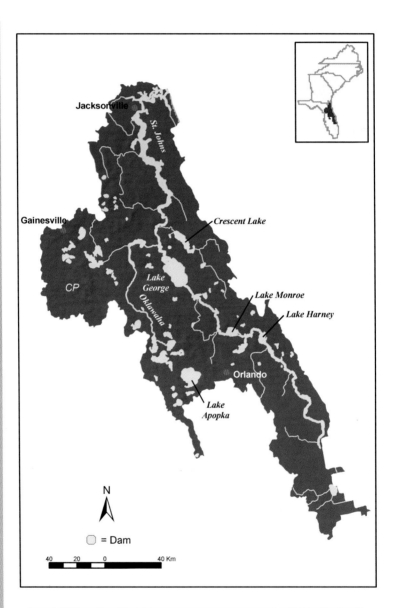

St. Johns River

Relief: 107 m
Basin area: 22,539 km²
Mean discharge: 222 m³/s
Mean annual precipitation: 131 cm

Mean air temperature: 20°C
Mean water temperature: 22°C
No. of fish species: >75 fresh-
water, 115 euryhaline
No. of endangered species: 9

Physiographic province: Coastal Plain (CP)

Major fishes: bluegill, redbreast sunfish, American eel, channel catfish, white catfish, largemouth bass, black crappie, American shad, striped bass

Major other aquatic vertebrates: cottonmouth, water snakes, snapping turtle, mud turtle, musk turtle, Florida cooter, bullfrog, treefrogs, American alligator, river otter, muskrat, beaver, West Indian manatee, brown pelican, herons, egrets, ibis

Major benthic insects: mayflies (*Callibaetis, Stenacron*), caddisflies (*Oecetis, Hydroptila, Orthotrichia, Cyrnellus*), true flies (*Chironomus, Glyptotendipes*)

Nonnative species: water hyacinth, hydrilla, Eurasian watermilfoil, common salvinia, parrot feather, Asiatic clam, nutria, numerous other plant and animal species

Major riparian plants: maidencane, pickerelweed, arrowhead, sawgrass, cattail, rush, cordgrass, coastal plain willow

Special features: channel–lake geomorphology; broad marsh floodplain; tidal influence far upriver; extremely low gradient

Fragmentation: no major dams on main stem, dam on Oklawaha River; altered hydroperiod due to water diversions

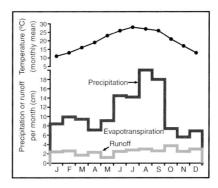

St. Johns River, north end of Lake George, Florida (photo by M. O'Malley/St. Johns River Water Management District).

York River

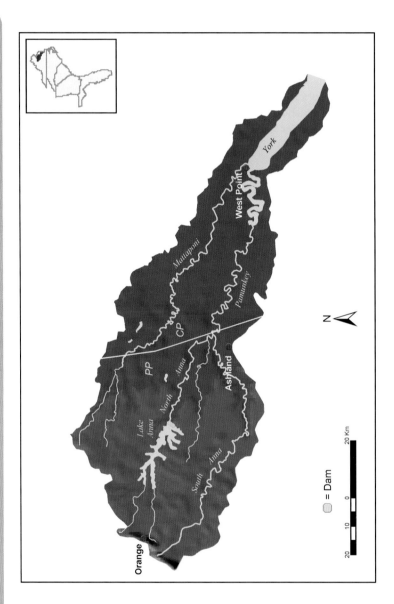

Relief: 365 m
Basin area: 6892 km^2
Mean discharge: 45 m^3/s
Mean annual precipitation: 108 cm

Mean air temperature: 14°C
Mean water temperature: 15°C
No. of fish species: 75
No. of endangered species: 7

Physiographic provinces: Piedmont Plateau (PP), Coastal Plain (CP)
Major fishes: longnose gar, American eel, American shad, blueback herring, pickerels, blue catfish, white perch, striped bass, bluegill, redbreast sunfish, largemouth bass, yellow perch
Major other aquatic vertebrates: northern water snake, brown water snake, painted turtle, musk turtle, mud turtle, red-bellied turtle, snapping turtle, muskrat, river otter, beaver
Major benthic insects: mayflies (*Baetis, Stenonema, Eurylophella, Caenis*), caddisflies (*Cheumatopsyche, Hydropsyche, Chimarra, Hydroptila*), true flies (*Rheotanytarsus*)
Nonnative species: common carp, grass carp, channel catfish, blue catfish, white crappie, black crappie, bluegill, smallmouth bass, largemouth bass, spotted bass, Asiatic clam

Major riparian plants: sycamore, swamp black gum, river birch, American elm, red maple, ash-leaf maple, bald cypress
Special features: strong anadromous fish runs and intact historical anadromous fish spawning grounds
Fragmentation: 1 major dam on primary tributary (North Anna River)

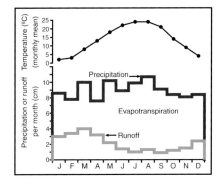

Mattaponi River (branch of York River) near West Point, Virginia (photo from Virginia Institute of Marine Science).

Roanoke River

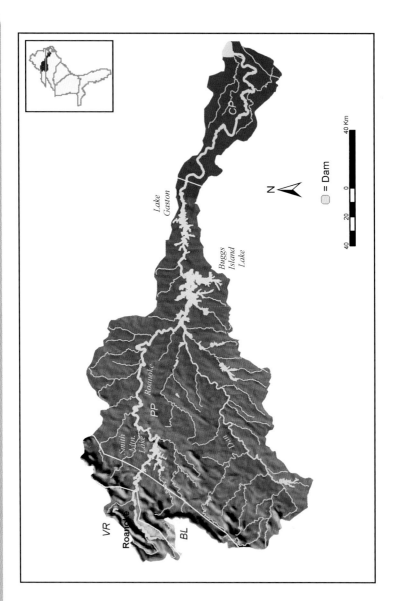

Relief: 920 m
Basin area: 25,326 km²
Mean discharge: 232 m³/s
Mean annual precipitation: 108 cm

Mean air temperature: 14°C
Mean water temperature: 16°C
No. of fish species: 119
No. of endangered species: 18

Physiographic provinces: Valley and Ridge (VR), Blue Ridge (BL), Piedmont Plateau (PP), Coastal Plain (CP)
Major fishes: American eel, American shad, hickory shad, blueback herring, alewife, gizzard shad, redhorses, shiners, darters, striped bass, white perch, redear sunfish, bluegill, small-mouth bass, largemouth bass, Roanoke bass, black crappie, yellow perch
Major other aquatic vertebrates: cottonmouth, water snakes, painted turtle, musk turtle, river cooter, slid-ers, mud turtle, snapping turtle, bull frog, muskrat, river otter, beaver
Major benthic insects: mayflies (*Stenonema, Baetis*), caddisflies (*Hydropsyche, Cheumatopsyche, Polycentropus*), true flies (*Rheotanytarsus*)
Nonnative species: Asiatic clam, smallmouth bass, rock bass, bluegill, flathead catfish, channel catfish, threadfin shad, walleye

Major riparian plants: sycamore, swamp black gum, American elm, red maple, ash-leaf maple, bald cypress, water tupelo, green ash, water ash, swamp chestnut oak
Special features: floodplain supports one of largest tracts of intact and largely undisturbed bottomland hardwood forests on Atlantic coast; 6 endemic species of fishes in the basin
Fragmentation: strongly fragmented by large dams on main stem

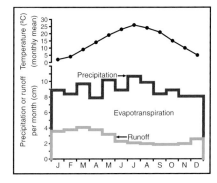

Roanoke River below Altavista, Virginia (photo by Tim Palmer).

Great Pee Dee River

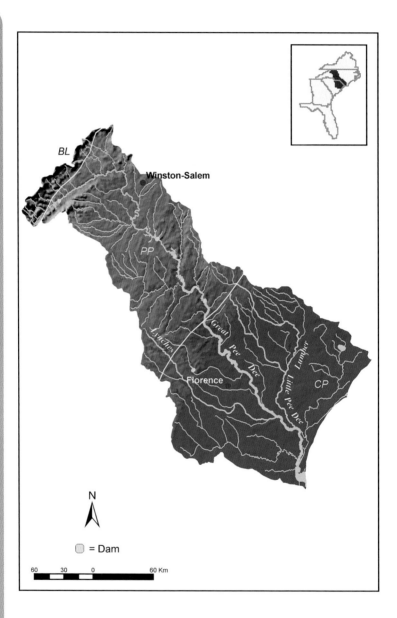

BL

Winston-Salem

PP

Lynches

Great

Pee

Dee

Lumber

Little Pee Dee

CP

Florence

N

⬤ = Dam

60   30   0   60 Km

Relief: 1090 m
Basin area: 27,560 km^2
Mean discharge: 371 m^3/s
Mean annual precipitation: 111 cm

Mean air temperature: 17°C
Mean water temperature: 17°C
No. of fish species: 101
No. of endangered species: 15

Physiographic provinces: Blue Ridge (BL), Piedmont Plateau (PP), Coastal Plain (CP)

Major fishes: bowfin, American eel, American shad, blueback herring, gizzard shad, common carp, eastern silvery minnow, shiners, silver redhorse, channel catfish, striped bass, white perch, redbreast sunfish, bluegill, redear sunfish, largemouth bass, black crappie

Major other aquatic vertebrates: cottonmouth, northern water snake, brown water snake, mud turtle, snapping turtle, musk turtle, river cooter, slider, river frog, bull frog, muskrat, river otter, beaver

Major benthic insects: mayflies (*Baetis*, *Caenis*, *Stenonema*, *Tricorythodes*), caddisflies (*Cheumatopsyche*, *Chimarra*, *Hydropsyche*, *Oecetis*), true flies (*Rheotanytarsus*, *Simulium*)

Nonnative species: Asiatic clam, flathead catfish, channel catfish

Major riparian plants: bald cypress, water tupelo, swamp black gum, sweetgum, water hickory, red maple, oaks

Special features: broad hardwood floodplain forests in Coastal Plain; Lumber River tributary is National Wild and Scenic River; Lumber and Lynches rivers are among few natural free-flowing rivers in coterminous 48 states

Fragmentation: several dams in Piedmont

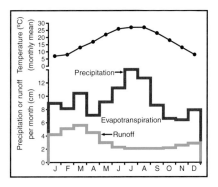

Great Pee Dee River, upstream of Route 32 near Brownsville, South Carolina (photo by A. C. Benke).

Santee River

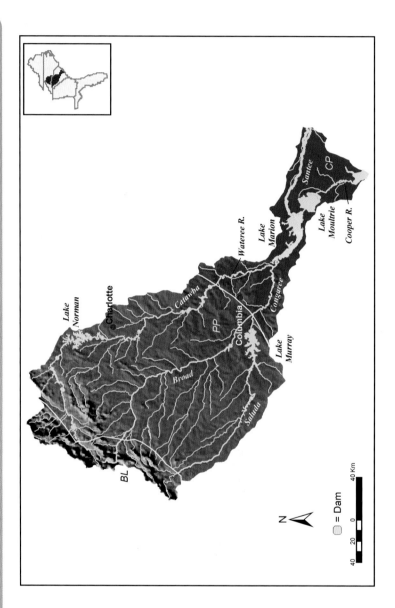

Relief: 1789 m
Basin area: 39,500 km²
Mean discharge: 434 m³/s
Mean annual precipitation: 125 cm

Mean air temperature: 17°C
Mean water temperature: 18°C
No. of fish species: 123
No. of endangered species: 7

Physiographic provinces: Blue Ridge (BL), Piedmont Plateau (PP), Coastal Plain (CP)

Major fishes: bowfin, American eel, American shad, blueback herring, gizzard shad, spotted sucker, shiners, channel catfish, striped bass, black crappie, redbreast sunfish, redear sunfish, bluegill, largemouth bass

Major other aquatic vertebrates: cottonmouth, water snakes, snapping turtle, mud turtle, musk turtle, river cooter, slider, river frog, bullfrog, muskrat, river otter, beaver

Major benthic insects: mayflies (*Baetis, Caenis, Stenonema, Tricorythodes*), caddisflies (*Cheumatopsyche, Chimarra, Hydropsyche, Oecetis*), true flies (*Rheotanytarsus, Simulium*)

Nonnative species: Asiatic clam, flathead catfish, channel catfish, white crappie, hydrilla

Major riparian plants: sycamore, bald cypress, water tupelo, swamp black gum, sweetgum, red maple, hickories, oaks

Special features: second largest river basin on Atlantic coast of United States

Fragmentation: many dams on main stem and major tributaries

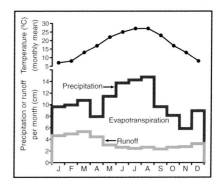

Congaree River (branch of Santee River), downstream of Congaree National Park, South Carolina (photo by A. C. Benke).

Altamaha River

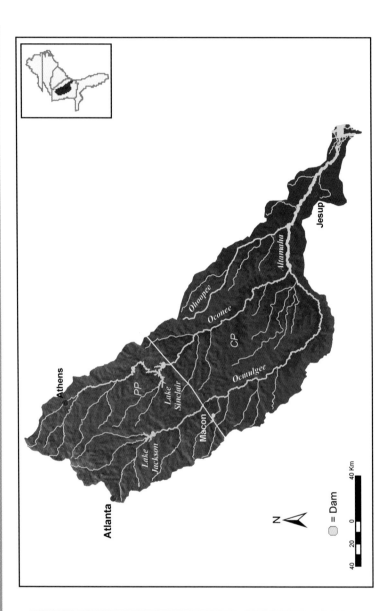

Relief: 372 m   Mean air temperature: 19°C
Basin area: 37,600 km²   Mean water temperature: 20°C
Mean discharge: 393 m³/s   No. of fish species: 93
Mean annual precipitation: 130 cm   No. of endangered species: 25

Physiographic provinces: Piedmont Plateau (PP), Coastal Plain (CP)
Major fishes: bowfin, American eel, American shad, gizzard shad, chain pickerel, carp, minnows, shiners, silver redhorse, spotted sucker, carpsucker, bullhead catfish, channel catfish, black crappie, bluegill, redear sunfish, warmouth, largemouth bass, hogchoker
Major other aquatic vertebrates: cottonmouth, water snakes, softshell turtles, river cooter, American alligator, river frog, sirens, treefrogs, dusky salamanders, West Indian manatee, muskrat, river otter, beaver
Major benthic insects: caddisflies (*Hydropsyche*, *Cheumatopsyche*, *Macrostemum*, *Chimarra*, *Neureclipsis*), mayflies (*Isonychia*, *Stenonema*, *Baetis*, *Tricorythodes*, *Caenis*), stoneflies (*Paragnetina*), true flies (*Polypedilium*, *Rheotanytarsus*, *Simulium*)
Nonnative species: Asiatic clam, flathead catfish
Major riparian plants: swamp black gum, water tupelo, bald cypress, water hickory, red maple, sweetgum, oaks
Special features: broad forested floodplain swamp; designated as a "Bioreserve" by the Nature Conservancy

Fragmentation: dams in Piedmont portions of two primary tributaries (Ocmulgee and Oconee rivers); no main-stem dams

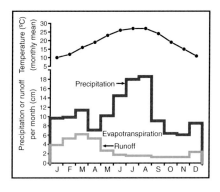

Altamaha River, about 4 km upstream of I-95 (photo by R. Overman).

Satilla River

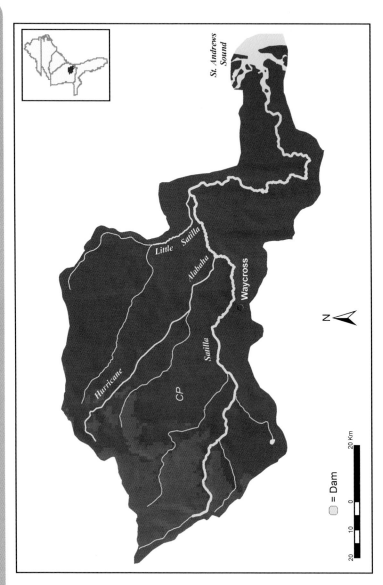

Relief: 107 m
Basin area: 9143 km^2
Mean discharge: 65 m^3/s
Mean annual precipitation: 126 cm

Mean air temperature: 19°C
Mean water temperature: 20°C
No. of fish species: 52
No. of endangered species: 6

Physiographic province: Coastal Plain (CP)
Major fishes: bowfin, American eel, chain pickerel, spotted sucker, channel catfish, yellow bullhead, black crappie, bluegill, redbreast sunfish, warmouth, largemouth bass, brook silverside, eastern mosquitofish, topminnow, pirate perch, spotted sunfish, minnows, darters
Major other aquatic vertebrates: cottonmouth, water snakes, softshell turtles, river cooter, loggerhead musk turtle, American alligator, bullfrog, treefrogs, dusky salamanders, river otter, muskrat, beaver, West Indian manatee
Major benthic insects: caddisflies (*Hydropsyche*, *Cheumatopsyche*, *Macrostemum*, *Chimarra*), stoneflies (*Perlesta*, *Acroneuria*), mayflies (*Stenonema*), true flies (*Simulium*, *Rheotanytarsus*, *Polypedilium*, *Rheosmittia*)

Nonnative species: flathead catfish
Major riparian plants: swamp black gum, water tupelo, bald cypress, water hickory, river birch, black willow, red maple
Special features: broad forested floodplain swamp; one of few natural free-flowing rivers in coterminous 48 states
Fragmentation: no dams on main stem; proposed titanium mining in floodplain near Atkinson

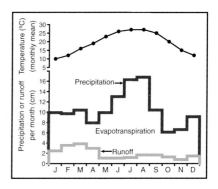

Satilla River, about 16 km downstream of Atkinson, Georgia (photo by A. C. Benke).

Chapter 4

Gulf Coast Rivers of the Southeastern United States

G. Milton Ward, Phillip M. Harris, and Amelia K. Ward

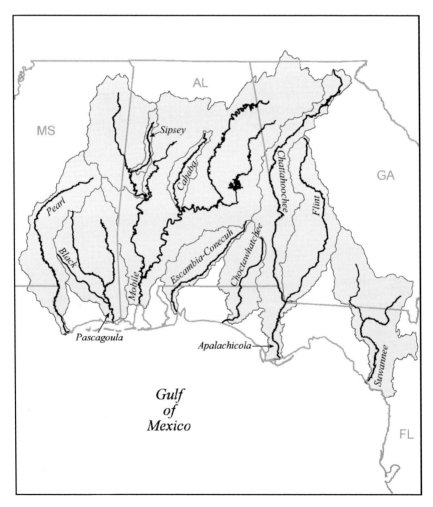

The river basins of the Eastern Gulf Coast lie west of the Atlantic slope and east of the Mississippi River. The climate of the region is moderate with warm summers, mild winters, and abundant rainfall. The region has abundant water resources, including seven major rivers that arise and flow through five physiographic provinces in five states to empty into the Gulf of Mexico (see map). Eastern Gulf Coast rivers encompass a rich variety of aquatic habitats and resources. In addition to the many upland streams and large rivers, there are large swamps, such as the Okefenokee Swamp and the Mobile River Delta, wide floodplain swamps, and oxbow lakes, which occur along all of the major rivers and many of the smaller coastal plain rivers. Limestone springs derived from the Floridan aquifer arise in southwestern Georgia, southeastern Alabama, and northern Florida. Today, the main stems of many major rivers of the Eastern Gulf are severely fragmented by numerous hydroelectric and navigation dams. Their mainstem channels have been deepened, and many riverine fauna have been replaced by lentic (reservoir) fauna. At the lower end of many basins are moderate to large estuaries, such as the Apalachicola, Mobile, Choctawhatchee, and Pearl, which are now, or were at one time, important centers of commercial and recreational fisheries.

Humans have inhabited these river basins for at least 12,000 years. Although hunter-gatherer societies occupied much of this time, the more sedentary lifestyle of the Mississippian culture dominated the southeastern region from 900 to 1500 A.D., reaching a peak between 1200 and 1400 A.D. By the time of the arrival of the first Europeans, the expansive Mississippian culture was already in decline.[1] The first European exploration was by the Spanish and later colonization was by the Spanish, French, and English, all who partnered with Native Americans to vie for control of important lands, especially commercial and military corridors such as rivers and bays. After the war with the French in 1756, the English gained official control after the Treaty of Paris ceded land east of the Mississippi to England in 1763.[2]

This chapter includes the largest rivers of the eastern Gulf drainage east of the Mississippi River: Suwannee, Apalachicola, Escambia-Conecuh, Choctawhatchee, Mobile, Pascagoula, and Pearl. The Mobile is by far the largest river with a basin area of $>111,000 \, km^2$ and a mean discharge of almost $2000 \, m^3/s$. The Apalachicola is the second largest at $>50,000 \, km^2$ and $759 \, m^3/s$. Features of smaller tributaries of the Mobile (Cahaba, Sipsey) and Apalachicola (Flint) are also described.

---

1 Scarry, J. F. (ed.). 1996. Political structure and change in the prehistoric Southeastern United States. University of Florida Press, Gainesville.
2 Jackson III, H. H. 1995. Rivers of History: Life on the Coosa, Tallapoosa, Cahaba, and Alabama. University of Alabama Press, Tuscaloosa.

Mobile River

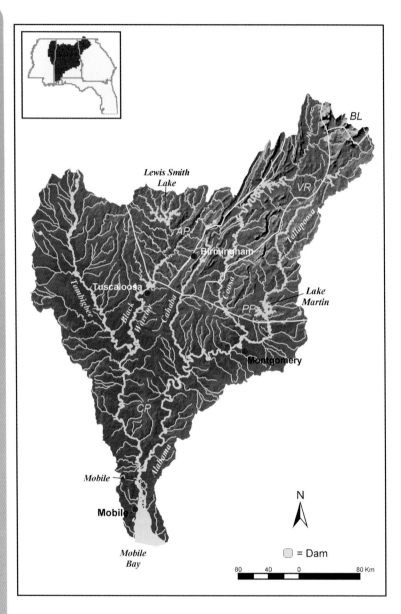

Relief: 1278 m
Basin area: 111,369 km^2
Mean discharge: 1914 m^3/s
Mean annual precipitation: 128 cm

Mean air temperature: 17.4°C
Mean water temperature: 19.9°C
No. of fish species: 236
No. of endangered species: 31

Physiographic provinces: Coastal Plain (CP), Valley and Ridge (VR), Appalachian Plateau (AP), Piedmont Plateau (PP), Blue Ridge (BL)

Major fishes: paddlefish, Gulf sturgeon, Alabama sturgeon, gars, shads, highfin carpsucker, southeastern blue sucker, spotted sucker, river redhorse, golden redhorse, channel catfish, flathead catfish, striped bass, green sunfish, bluegill, longear sunfish, spotted bass, largemouth bass, white crappie, black crappie, freshwater drum

Major other aquatic vertebrates: American alligator, cottonmouth, diamondback water snake, alligator snapping turtle, common snapping turtle, flattened musk turtle, Alabama map turtle, southern painted turtle, Alabama redbelly turtle

Major benthic insects: mayflies (*Stenacron, Stenonema, Isonychia, Baetis*), stoneflies (*Acroneuria, Perlesta*), caddisflies (*Hydropsyche, Cheumatopsyche, Hydroptila, Ceraclea, Oecetis*)

Nonnative species: common carp, silver carp, bighead carp, grass carp, muskellunge, rainbow trout, brown trout, smallmouth bass, goldfish

Major riparian plants: bald cypress, river birch, American beech, southern red oak, water oak, live oak, yellow poplar, sweetgum, American sycamore, red maple, black gum, water tupelo, swamp tupelo

Special features: largest flow into eastern Gulf of Mexico; high diversity of fishes, turtles, mollusks, insects; high levels of endemism and high levels of threatened/endangered species; large pristine forest at mouth

Fragmentation: 36 major dams

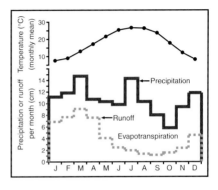

Tensaw Delta, a distributary of the lower Mobile River, northeast of Mobile, Alabama (photo by Beth Maynor Young).

Cahaba River

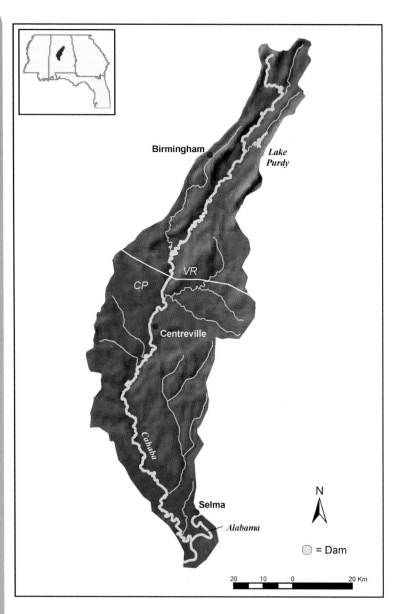

Birmingham

Lake Purdy

VR

CP

Centreville

Cahaba

Selma

Alabama

N

◯ = Dam

20   10   0   20 Km

Relief: 274 m
Basin area: 4730 km^2
Mean discharge: 80 m^3/s
Mean annual precipitation: 138 cm

Mean air temperature: 16.7°C
Mean water temperature: 18.1°C
No. of fish species: 135
No. of endangered species: 7

Physiographic provinces: Valley and Ridge (VR), Coastal Plain (CP)

Major fishes: paddlefish, Alabama sturgeon, gars, shads, blacktail shiner, quillback, highfin carpsucker, southeastern blue sucker, spotted sucker, river redhorse, golden redhorse, blacktail redhorse, blue catfish, channel catfish, flathead catfish, white bass, warmouth, green sunfish, bluegill, longear sunfish, spotted bass, largemouth bass, crappie

Major other aquatic vertebrates: American alligator, alligator snapping turtle, common snapping turtle, Alabama map turtle, Gulf Coast smooth softshell turtle, southern painted turtle, cottonmouth, diamondback water snake

Major benthic insects: mayflies (*Stenacron, Stenonema, Tricorythodes, Caenis*), stoneflies (*Acroneuria, Perlesta*), caddisflies (*Cheumatopsyche, Hydropsyche, Hydroptila, Cyrnellus, Ceraclea*)

Nonnative species: grass carp, common carp, fathead minnow, white catfish, palmetto bass, smallmouth bass, Asiatic clam

Major riparian plants: bald cypress, river birch, American beech, southern red oak, water oak, live oak, yellow poplar, sweetgum, American sycamore, red maple, black gum, water tupelo, swamp tupelo

Special features: longest free-flowing river in southeastern Gulf Coast region; most fish species for its size in North America; very high mollusk diversity

Fragmentation: small impoundment in headwater tributary

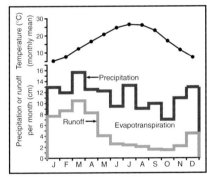

Cahaba River at "Lily Shoals" southwest of Birmingham, Alabama (photo by Beth Maynor Young).

Apalachicola–Chattahoochee–Flint River

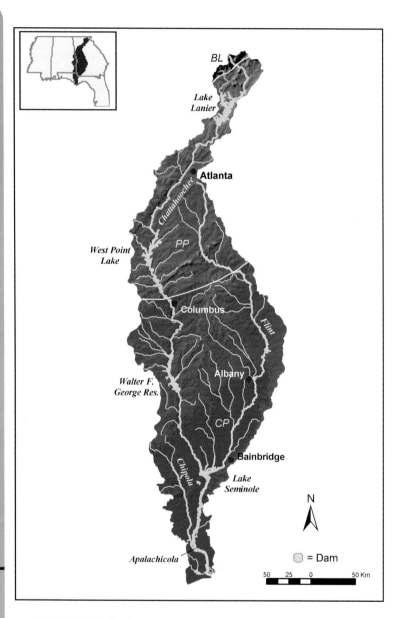

BL

Lake
Lanier

Chattahoochee

Atlanta

*West Point
Lake*

PP

Columbus

Flint

*Walter F.
George Res.*

Albany

CP

Bainbridge

Chipola

*Lake
Seminole*

N

◯ = Dam

50 25 0   50 Km

Apalachicola

Relief: 1066 m

Basin area: 50,688 km^2

Mean discharge: 759 m^3/s

Mean annual precipitation: 128 cm

Mean air temperature: 18.3°C

Mean water temperature: 20.6°C

No. of fish species: 104

No. of endangered species: 2

Physiographic provinces: Blue Ridge (BL), Piedmont Plateau (PP), Coastal Plain (CP)

Major fishes: Gulf sturgeon, longnose gar, American eel, Alabama shad, gizzard shad, Apalachicola redhorse, channel catfish, white catfish, yellow bullhead, brown bullhead, spotted bullhead, striped bass, largemouth bass, shoal bass, black crappie, redbreast sunfish, warmouth, bluegill

Major other aquatic vertebrates: American alligator, cottonmouth, alligator snapping turtle, Barbor's map turtle, Florida softshell turtle, Gulf Coast spiny softshell turtle, Florida redbelly turtle, Florida cooter, Alabama waterdog

Major benthic insects: mayflies (*Hexagenia, Stenonema, Baetis, Caenis*), stoneflies (*Acroneuria, Paragnetina, Perlesta*), caddisflies (*Ceraclea, Cheumatopsyche, Hydropsyche, Oecetis*)

Nonnative species: common carp, grass carp, green sunfish, orange spotted sunfish, walking catfish, goldfish, tilapia, Asian swamp eel, Asiatic clam

Major riparian plants: water tupelo, Ogeechee tupelo, bald cypress, swamp tupelo, sweetgum, overcup oak, water hickory, water oak, red maple, sweetbay

Special features: second largest drainage into eastern Gulf of Mexico, high diversity of channel gradients and geology; historically diverse mollusk, crayfish, and fish fauna

Fragmentation: highly fragmented, with 16 main-stem dams, 13 on Chattahoochee River and 3 on Flint River

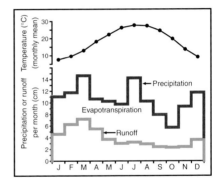

Chattahoochee River (Apalachicola system) below Buford Dam, north of Atlanta (photo by A. C. Benke).

Pearl River

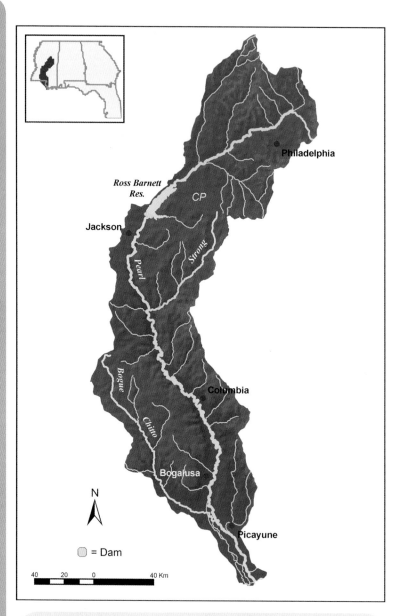

Relief: 210 m

Basin area: 21,999 km^2

Mean discharge: 373 m^3/s

Mean annual precipitation: 142 cm

Mean air temperature: 17.8°C

Mean water temperature: 19.2°C

No. of fish species: 119

No. of endangered species: 4

Physiographic province: Coastal Plain (CP)

Major fishes: paddlefish, Gulf sturgeon, alligator gar, gizzard shad, highfin carpsucker, southeastern blue sucker, smallmouth buffalo, blacktail redhorse, yellow bullhead, channel catfish, flathead catfish, warmouth, green sunfish, bluegill, longear sunfish, redspotted sunfish, spotted bass, largemouth bass, white crappie, black crappie, drum

Major other aquatic vertebrates: river otter, American alligator, alligator snapping turtle, common snapping turtle, stripeneck musk turtle, Pascagoula map turtle, ringed map turtle, Gulf Coast spiny softshell turtle, cottonmouth

Major benthic insects: mayflies (*Stenonema, Baetis, Caenis, Tricorythodes, Isonychia*), stoneflies (*Paragnetina, Acroneuria, Neoperla*), caddisflies (*Hydropsyche, Cheumatopsyche, Hydroptila, Chimarra, Ceraclea*)

Nonnative species: fathead minnow, common carp, goldfish, Asiatic clam

Major riparian plants: eastern cottonwood, black willow, river birch, sycamore, silver maple, laurel oak, willow oak, water oak, sugarberry, American elm, green ash, overcup oak, water hickory, bald cypress, pond cypress, black gum, water hickory oak, water tupelo, sweetbay, red maple, slash pine

Special features: low-gradient Coastal Plain river with swamps and wide floodplain forests; large coastal estuary; high fish species richness

Fragmentation: one large impoundment, Ross Barnett dam, near Jackson, Mississippi

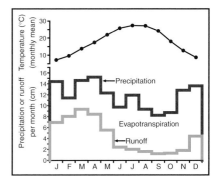

Pearl River near Slidell, Louisiana (photo by G. M. Ward).

Suwannee River

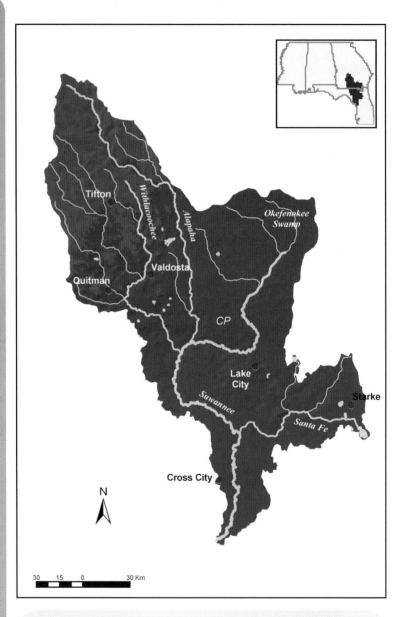

Relief: 140 m
Basin area: 24,967 km²
Mean discharge: 294 m³/s
Mean annual precipitation: 134 cm

Mean air temperature: 20.2°C
Mean water temperature: 19.7°C
No. of fish species: 81
No. of endangered species: 1

Physiographic province: Coastal Plain (CP)

Major fishes: Gulf sturgeon, longnose gar, Florida gar, gizzard shad, threadfin shad, blacktail shiner, spotted sucker, channel catfish, white catfish, flathead catfish, snail bullhead, yellow bullhead, brown bullhead, spotted bullhead, striped bass, Suwannee bass, largemouth bass, black crappie, warmouth, redbreast sunfish, spotted sunfish, bluegill

Major other aquatic vertebrates: river otter, West Indian manatee, American alligator, alligator snapping turtle, Florida snapping turtle, Florida softshell turtle, Florida redbelly turtle, Suwannee cooter, Florida cooter, Florida cottonmouth, Gulf saltmarsh swamp snake, North Florida swamp snake, redbelly water snake, banded water snake, Florida green water snake, brown water snake

Major benthic insects: mayflies (*Baetis, Pseudocloeon, Stenacron, Stenonema, Tricorythodes*), stoneflies (*Acroneuria, Neoperla, Attaneuria, Paragnetina, Perlesta*), caddisflies (*Cheumatopsyche, Hydropsyche, Hydroptila, Ceraclea, Oecetis, Chimarra*)

Nonnative species: American shad, grass carp, blue catfish, flathead catfish, wiper, Asiatic clam

Major riparian plants: bald cypress, water elm, swamp laurel oak, overcup oak, live oak, sand live oak, sweetgum, river birch, planer tree, cabbage palm, red maple, water tupelo

Special features: large free-flowing Coastal Plain river with headwater swamps (Okefenokee Swamp); swamps and wide floodplain forests in mid and lower reaches; large hydrologic inputs from many springs flowing from underlying Floridan aquifer

Fragmentation: no impoundments

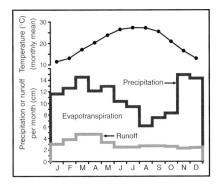

Lower Suwannee River (photo by Tim Palmer).

Choctawhatchee River

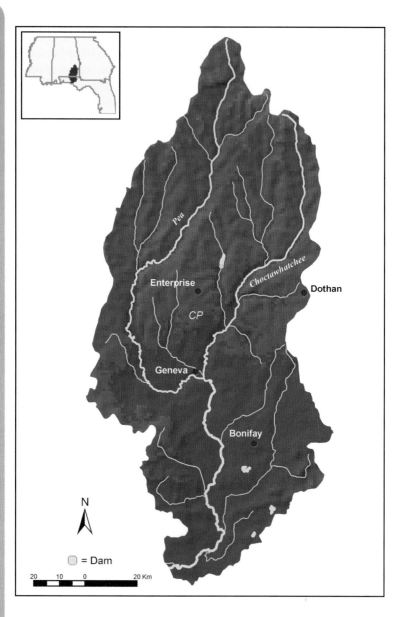

Relief: 179 m
Basin area: 12,033 km²
Mean discharge: 212 m³/s
Mean annual precipitation: 144 cm

Mean air temperature: 18.7°C
Mean water temperature: 20.0°C
No. of fish species: 80
No. of endangered species: 1

Physiographic province: Coastal Plain (CP)

Major fishes: Gulf sturgeon, spotted gar, longnose gar, American eel, Alabama shad, gizzard shad, threadfin shad, blacktail shiner, quillback, highfin carpsucker, blacktail redhorse, yellow bullhead, brown bullhead, channel catfish, striped bass, warmouth, bluegill, green sunfish, longear sunfish, spotted bass, largemouth bass, black crappie

Major other aquatic vertebrates: American alligator, Alabama waterdog, cottonmouth, alligator snapping turtle, Florida softshell turtle, Gulf Coast spiny softshell turtle, common snapping turtle, yellowbelly turtle, Florida cooter

Major benthic insects: mayflies (*Isonychia, Leptophlebia, Eurylophella*), stoneflies (*Neoperla, Paragnetina, Perlesta, Clioperla, Isoperla*), caddisflies (*Cheumatopsyche, Hydropsyche, Hydroptila, Oecetis*)

Nonnative species: grass carp, common carp, palmetto bass, yellow perch, unidentified pacu, Asiatic clam

Major riparian plants: bald cypress, black gum, water tupelo, swamp tupelo, Ogeechee tupelo, sweetgum, red maple, river birch, Atlantic white cedar, eastern cottonwood, swamp cottonwood, water hickory, American beech, water oak, live oak, yellow poplar, American sycamore

Special features: low-gradient Coastal Plain river with swamps and wide floodplain forest; one of few free-flowing southeastern Gulf of Mexico rivers; flows through lightly populated landscape into commercially valuable estuary

Fragmentation: no impoundments

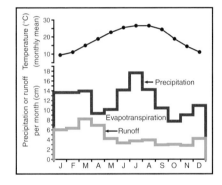

Choctawhatchee River, west of Dothan, Alabama (photo by Beth Maynor Young).

Escambia–Conecuh River

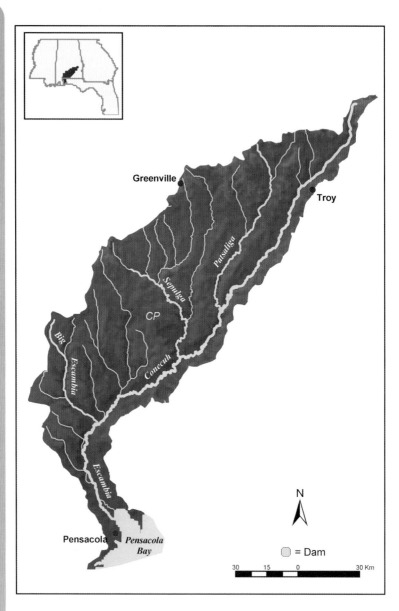

Relief: 180 m
Basin area: 10,963 km²
Mean discharge: 196 m³/s
Mean annual precipitation: 164 cm

Mean air temperature: 18.0°C
Mean water temperature: 20.4°C
No. of fish species: 102
No. of endangered species: 0

Physiographic province: Coastal Plain (CP)

Major fishes: Gulf sturgeon, spotted gar, longnose gar, Alabama shad, gizzard shad, threadfin shad, blacktail shiner, quillback, highfin carpsucker, river redhorse, blacktail redhorse, yellow bullhead, brown bullhead, channel catfish, flathead catfish, striped bass, warmouth, bluegill, green sunfish, longear sunfish, spotted bass, largemouth bass, black crappie

Major other aquatic vertebrates: American alligator, Alabama waterdog, cottonmouth, alligator snapping turtle, Florida softshell turtle, Gulf Coast spiny softshell turtle, common snapping turtle, Escambia map turtle, Florida cooter, stinkpot

Major benthic insects: mayflies (*Baetis, Isonychia, Leptophlebia, Hexagenia, Eurylophella*), stoneflies (*Acroneuria, Paragnetina, Clioperla, Isoperla*), caddisflies (*Ceraclea, Cheumatopsyche, Hydropsyche, Hydroptila, Oecetis, Oxyethira*)

Nonnative species: grass carp, common carp, palmetto bass

Major riparian plants: bald cypress, river birch, American beech, southern red oak, water oak, live oak, yellow poplar, sweetgum, American sycamore, red maple, black gum, water tupelo, swamp tupelo

Special features: low-gradient Coastal Plain river with swamps and wide floodplain forests in lower reaches; empties into a once productive estuary

Fragmentation: largely free flowing, although two moderate-size impoundments on main stem

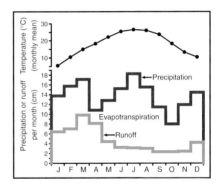

Escambia River upstream from Pensacola (photo by Beth Maynor Young).

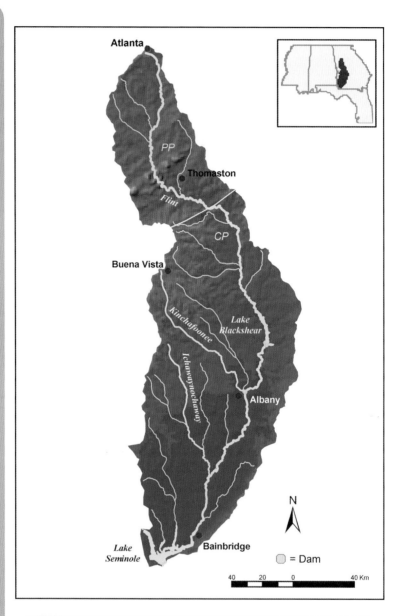

Flint River

Relief: 294 m   Mean air temperature: 18.2°C
Basin area: 22,377 km²   Mean water temperature: 19.8°C
Mean discharge: 283 m³/s   No. of fish species: 71
Mean annual precipitation: 126 cm No. of endangered species: 5

Physiographic provinces: Coastal Plain (CP), Piedmont Plateau (PP)

Major fishes: Gulf sturgeon, longnose gar, American eel, Alabama shad, gizzard shad, spotted sucker, Apalachicola redhorse, greater jumprock, channel catfish, white catfish, snail bullhead, brown bullhead, spotted bullhead, yellow bullhead, striped bass, black crappie, shoal bass, largemouth bass, warmouth, redbreast sunfish, bluegill

Major other aquatic vertebrates: river otter, American alligator, Alabama waterdog, redbelly water snake, brown water snake, cottonmouth, snapping turtle, alligator snapping turtle, Barbor's map turtle, Florida redbelly turtle, striped mud turtle, common musk turtle, Florida softshell turtle, Gulf Coast spiny softshell turtle, Florida cooter, river cooter, yellowbelly slider

Major benthic insects: mayflies (*Stenonema*, *Tricorythodes*, *Baetis*), stoneflies (*Paragnetina*), caddisflies (*Hydropsyche*, *Cheumatopsyche*, *Cyrnellus*)

Nonnative species: common carp, grass carp, flathead catfish, green sunfish, orange spotted sunfish, spotted bass, striped bass, goldfish, tilapia, walking catfish, Asian swamp eel

Major riparian plants: spruce pine, eastern hemlock, river birch, American hornbeam, American beech, white oak, water oak, laurel oak, American elm, sugarberry, umbrella magnolia, sweetbay, yellow poplar, sweetgum, American sycamore, red maple, box elder, water tupelo, black gum

Special features: second largest tributary to Apalachicola; karst terrain in lower reaches; habitat for Gulf sturgeon

Fragmentation: three dams, two large reservoirs on main stem for hydroelectric production

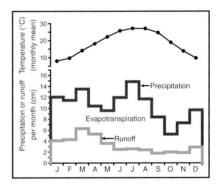

Flint River near Thomaston, Georgia, west of Macon (photo by Beth Maynor Young).

Pascagoula River–Black Creek

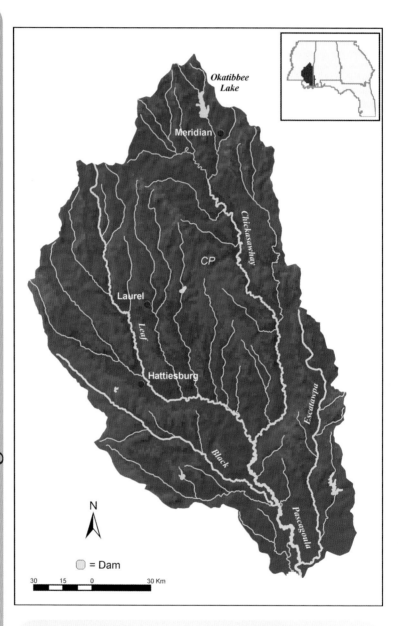

Relief: 198 m
Basin area: 24,599 km²
Mean discharge; 432 m³/s
Mean annual precipitation: 156 cm

Mean air temperature: 18.7°C
Mean water temperature: 19.7°C
No. of fish species: 114
No. of endangered species: 3

Physiographic province: Coastal Plain (CP)

Major fishes: paddlefish, Gulf sturgeon, spotted gar, longnose gar, gizzard shad, highfin carpsucker, southeastern blue sucker, smallmouth buffalo, blacktail redhorse, yellow bullhead, channel catfish, flathead catfish, shadow bass, warmouth, green sunfish, bluegill, redspotted sunfish, spotted bass, largemouth bass, white crappie, black crappie, freshwater drum

Major other aquatic vertebrates: beaver, mudpuppy, cottonmouth, American alligator, alligator snapping turtle, Pascagoula map turtle, Gulf Coast spiny softshell turtle

Major benthic insects: mayflies (*Stenonema, Baetis, Caenis, Tricorythodes, Isonychia*), stoneflies (*Paragnetina, Neoperla, Acroneuria*), caddisflies (*Hydropsyche, Cheumatopsyche, Hydroptila, Chimarra, Ceraclea*)

Nonnative species: bigheaded carp, fathead minnow, pirapitinga, Tilapia spp.

Major riparian plants: river birch, sycamore, laurel oak, willow oak, water oak, water hickory, bald cypress, pond cypress, black gum, water hickory oak, water tupelo, sweetbay, red maple

Special features: low-gradient blackwater ecosystem with swamps and wide floodplain forests; free flowing; 34 km reach of Black Creek classified as Wild and Scenic

Fragmentation: no dams on main stem or major tributaries

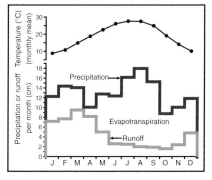

Pascagoula River: Oxbow Lake near confluence of Chickasawhay and Leaf rivers (photo by Beth Maynor Young).

Sipsey River

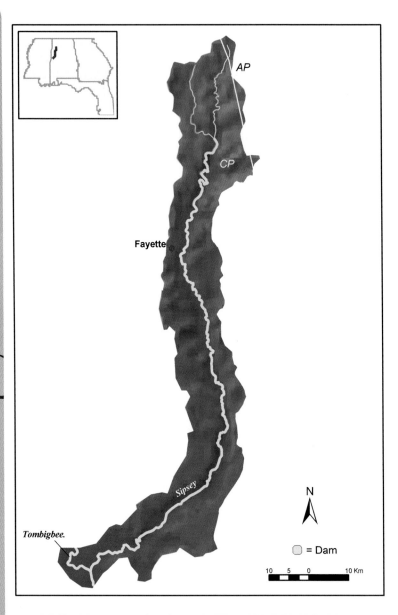

Relief: 229 m
Basin area: 2044 km^2
Mean discharge: 34 m^3/s
Mean annual precipitation: 139 cm

Mean air temperature: 17.0°C
Mean water temperature: 17.1°C
No. of fish species: 88
No. of endangered species: 0

Physiographic provinces: Coastal Plain (CP), Appalachian Plateau (AP)
Major fishes: spotted gar, longnose gar, gizzard shad, threadfin shad, blacktail shiner, quillback, smallmouth buffalo, Alabama hog sucker, spotted sucker, river redhorse, golden redhorse, blacktail redhorse, channel catfish, warmouth, bluegill, longear sunfish, redear sunfish, redspotted sunfish, spotted bass, largemouth bass, freshwater drum
Major other aquatic vertebrates: river otter, cottonmouth, yellowbelly water snake, diamondback water snake, American alligator, alligator snapping turtle, common snapping turtle, common musk turtle, stripeneck musk turtle, northern black-knob sawback map turtle, eastern chicken turtle, southern painted turtle, Gulf Coast smooth softshell turtle, spiny softshell turtle, river cooter, red-eared slider
Major benthic insects: mayflies (*Isonychia, Eurylophella, Serratella, Baetis, Caenis*), stoneflies (*Acroneuria, Perlesta, Isoperla, Neoperla, Taeniopteryx*), caddisflies (*Cheumatopsyche, Hydropsyche, Hydroptila, Ceraclea, Oecetis, Chimarra*)
Nonnative species: common carp
Major riparian plants: bald cypress, eastern cottonwood, swamp cottonwood, mockernut hickory, river birch, American hornbeam, American beech, southern red oak, water oak, live oak, American elm, yellow poplar, sweetgum, American sycamore, American holly, red maple, black gum, water tupelo, swamp tupelo, Carolina ash
Special features: lightly populated low-gradient Coastal Plain river with swamps and wide floodplain forests; high diversity of fish and mollusks
Fragmentation: no impoundments

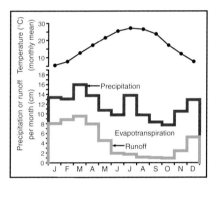

Sipsey River near Elrod, Alabama, west of Tuscaloosa (photo by A. C. Benke).

Chapter 5

Gulf Coast Rivers of the Southwestern United States

Clifford N. Dahm, Robert J. Edwards, and Frances P. Gelwick

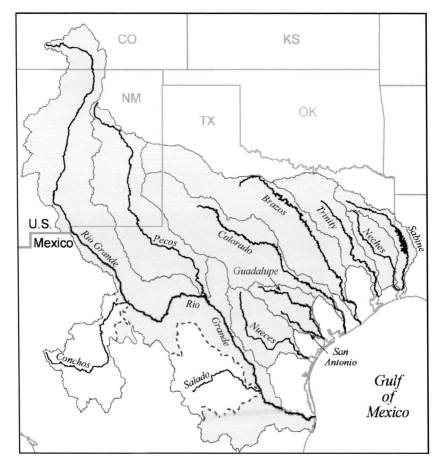

DOI: 10.1016/B978-0-12-375088-4.00005-1

Rivers flowing into the Texas portion of the Gulf of Mexico encompass a broad geographic area with latitude ranging from around 38° N in southern Colorado to 25° N in northern Mexico and with longitude ranging from about 108° W in western New Mexico to 93° W in western Louisiana (see map). The northwestern portion includes the southern Rocky Mountains while the southeast includes the flat coastal areas. The region is traversed by a strong decreasing rainfall gradient from east to west and a temperature gradient from north to south that strongly influences vegetation, land use, and river flow. These rivers encounter increasingly arid conditions as they flow southward through grasslands, shrublands, and deserts. Northeastern rivers of the western Gulf flow through prairies, pine forests, and cypress-lined bayous. Rivers crossing the desert and semiarid western parts of the region produce considerably less discharge to the Gulf of Mexico than the rivers of east Texas.

Human habitation within catchments that flow into the western Gulf of Mexico dates back to the Clovis culture of Paleo-Indians nearly 12,000 years ago. These big-game hunters are the earliest definitively dated human populations in the Americas. The Folsom culture also was an early hunting group of Paleo-Indians in the region that were present up until about 2500 years ago. Highly developed Native American civilizations became established one to two thousand years ago in both the eastern and western areas of rivers flowing into the western Gulf of Mexico, including the Mississippian mound-building culture in the Sabine basin and the Anasazi culture in the Rio Grande and Colorado basins. Early Spanish explorers visited the region in the sixteenth century.

Several major rivers discharge into the western Gulf of Mexico from the United States, including the Rio Grande, which borders with Mexico. They are located in generally narrow catchments that are from two to five times longer than their average widths and they have generally developed dendritic drainage systems. Most of these rivers are covered in this chapter, and include the Rio Grande, Nueces, San Antonio/Guadalupe, Colorado, Brazos, Trinity, Neches, and Sabine (see map). Also included is the Pecos, a major tributary of the Rio Grande. Many of these rivers are of great length with the Rio Grande, Brazos, Pecos, and Colorado among the hundred longest rivers in the world.[1] The largest river basin is the Rio Grande (870,000 km^2) but its mean discharge is currently only 40 m^3/s or less. The Colorado and Brazos are the only other basins exceeding >100,000 km^2, and the Brazos has the highest discharge of 249 m^3/s.

---

1 World Almanac Books. 2003. *The World Almanac and Book of Facts*. New York.

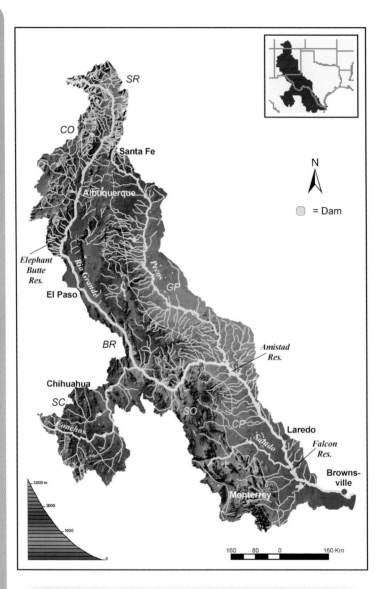

Rio Grande

Relief: 4272 m
Basin area: 870,000 km^2
(~450,000 km^2)
Mean discharge: 37 m^3/s (virgin
>100 m^3/s)
Mean annu. precipitation: 21 cm

Mean air temperature: 13°C
Mean water temperature: 14°C
No. of fish species: ≥86
freshwater, ≥80 estuarine
No. of endangered species:
>25

Physiographic provinces: Southern Rocky Mountains (SR), Colorado Plateau (CO), Basin and Range (BR), Great Plains (GP), Coastal Plain (CP), Sierra Madre Occidental (SC), Sierra Madre Oriental (SO)

Major fishes: Rio Grande cutthroat trout, red shiner, Rio Grande silvery minnow, fathead minnow, white sucker, blue sucker, river carpsucker, western mosquitofish, largemouth bass, bluegill, longnose gar, threadfin shad, Rio Grande shiner, Tamaulipas shiner, longnose dace, Mexican tetra, sailfin molly, Amazon molly, longear sunfish, Rio Grande cichlid

Major other aquatic vertebrates: common yellow-throat, great blue heron, snowy egret, black-crowned night heron, white-faced ibis, belted kingfisher, green kingfisher, plainbelly water snake, American alligator, beaver, mink, nutria, bullfrog, Rio Grande leopard frog, snapping turtle, painted turtle, box turtle, western ribbon snake

Major benthic insects: caddisflies (*Cheumatopsyche*, *Brachycentrus*, *Leucotrichia*, *Stactobiella*, *Hydroptila*, *Protoptila*), mayflies (*Baetis*, *Tricorythodes*, *Thraulodes*, *Traverella*, *Choroterpes*), true flies (*Cricotopus*, *Orthocladius*, *Atherix*, *Simulium*)

Nonnative species: ≥23 fishes (common carp, blue tilapia, inland silversides); saltcedar, Russian olive, Siberian elm, white mulberry, Guinea grass, buffelgrass, water hyacinth, hydrilla, water lettuce, alligatorweed, Asiatic clam, nutria

Major riparian plants: cottonwoods, willows, saltcedar, Russian olive, mesquite, hackberry, cedar elm, anacua, black willow, retama, Guinea grass, buffelgrass

Special features: Bosque del Apache National Wildlife Refuge, Chamizal National Memorial, Big Bend National Park, National Wild and Scenic River designation for segments in New Mexico and Texas

Fragmentation: highly fragmented, with main-stem and numerous tributary dams

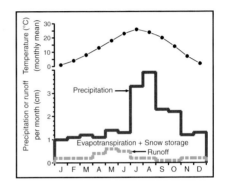

Rio Grande at Big Bend National Park, western Texas (photo by A. D. Huryn).

San Antonio and Guadalupe Rivers

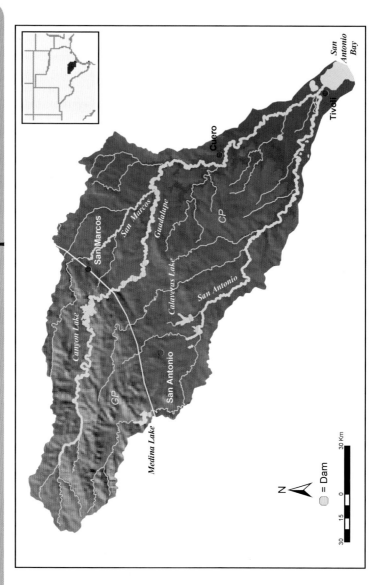

Relief: 700 m
Basin area: 26,231 km²
Mean discharge: 79 m³/s
Mean precipitation: 81 cm

Mean air temp.: 21°C
Mean water temp.: 23°C
No. of fish species: ≥88 (60 native)
No. of endangered species: >11

Physiographic provinces: Great Plains (GP), Coastal Plain (CP)
Major fishes: largemouth bass, Guadalupe bass, bluegill, longear sunfish, redear sunfish, spotted sunfish, blacktail shiner, red shiner, central stonerollers, gray redhorse, channel catfish, western mosquitofish, Texas shiner, Texas logperch, river carpsucker, smallmouth buffalo, spotted bass, dusky darter, orangethroat darter, greenthroat darter, bluntnose darter, river darter
Major other aquatic vertebrates: beaver, northern parula warbler, prothonotary warbler, Louisiana waterthrush, great blue heron, snowy egret, white-faced ibis, belted kingfisher, green kingfisher, American alligator, Texas map turtle, Cagle's map turtle, smooth softshell turtle, spiny softshell turtle, plainbelly water snake, Western cottonmouth, nutria, mink
Major benthic insects: caddisflies (*Chimarra, Cheumatopsyche, Atopsyche, Hydroptila*), mayflies (*Dactylobaetis, Tricorythodes, Choroterpes, Thraulodes*), true flies (*Cricotopus, Rheotanytarsus*)
Nonnative species: at least 27 fishes (tilapia, crappie, walleye, smallmouth bass, goldfish, grass carp), nutria, Asiatic clam, ramshorn snail, elephant ears, alligatorweed, water hyacinth, hydrilla, water lettuce
Major riparian plants: pecan, Texas sugarberry, bald cypress, cedar elm, Virginia creeper, Texas persimmon, red mulberry, greenbrier, box

elder, black walnut, cottonwood, gum bumelia, black willow, American elm
Special features: large artesian aquifers (associated with Edwards aquifer), San Marcos spring
Fragmentation: major dam on Upper Guadalupe (Canyon); smaller dams on San Antonio

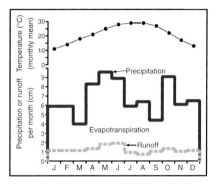

Guadalupe River above Rt. 311 (photo by Tim Palmer).

Colorado River

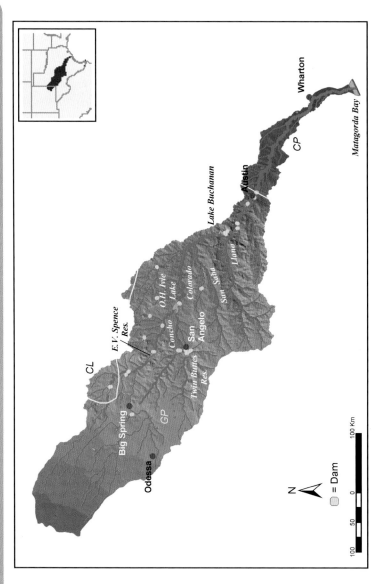

Relief: 1195 m
Basin area: 103,341 km²
Mean discharge: 75 m³/s
Mean annu. precipitation: 82 cm

Mean air temp.: 20°C
Mean water temp.: 22°C
No. of fish species: >98
(72 native)
No. of endangered species: >12

Physiographic provinces: Central Lowland (CL), Great Plains (GP), Coastal Plain (CP)

Major fishes: spotted gar, longnose gar, red shiner, bullhead minnow, central stoneroller minnow, suckermouth minnow, blue sucker, gray redhorse, yellow bullhead, channel catfish, blue catfish, western mosquitofish, blackstripe topminnow, longear sunfish, spotted bass, Guadalupe bass, orangethroat darter, greenthroat darter, Texas logperch, bigscale logperch

Major other aquatic vertebrates: beaver, great blue heron, snowy egret, white-faced ibis, belted kingfisher, green kingfisher, Texas map turtle, Texas River cooter, red-eared slider, smooth softshell turtle, spiny softshell turtle, American alligator, plainbelly water snake, Concho water snake, Harter's water snake, western cottonmouth, gray treefrog, cricket frog, southern leopard frog

Major benthic insects: caddisflies (*Cheumatopsyche*, *Helicopsyche*, *Hydroptila*), mayflies (*Tricorythodes*, *Caenis*), true flies (*Rheotanytarsus*, *Tanytarsus*)

Nonnative species: common carp, grass carp, rudd, walleye, northern pike, nutria, saltcedar, water hyacinth, hydrilla, water lettuce, giant salvinia, Eurasian watermilfoil, alligatorweed

Major riparian plants: live oak, red oak, sugarberry, sycamore, elm, black willow, eastern cottonwood, pecan

Special features: three reaches are Texas Natural Rivers System candidates by U.S. Park Service; 18-m Gorman Falls upstream of Lake Buchanan; class IV rapids at Crabapple Creek; 15-m falls into collapsed grotto at Hamilton Pool Preserve

Fragmentation: most heavily dammed river in Texas (25 hydroelectric and water-supply reservoirs)

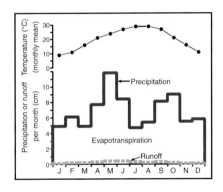

Colorado River at Wharton, Texas (photo by Tim Palmer).

Brazos River

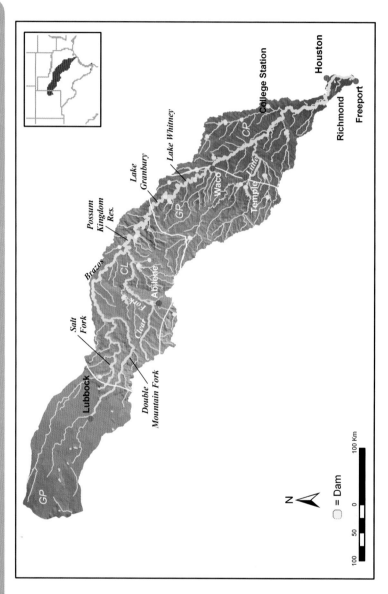

Relief: 1204 m
Basin area: 115,566 km²
Mean discharge: 249 m³/s
Mean annu. precipitation: 81 cm

Mean air temp.: 19°C
Mean water temp.: 21°C
No. of fish species: >93 (≥72 native)
No. of endangered species: >8

Physiographic provinces: Central Lowland (CL), Great Plains (GP), Coastal Plain (CP)

Major fishes: longnose gar, spotted gar, western mosquitofish, blackstripe topminnow, red shiner, blacktail shiner, ribbon shiner, pugnose minnow, bullhead minnow, bluegill, green sunfish, longear sunfish, white crappie, largemouth bass, spotted bass, slough darter, orangethroat darter

Major other aquatic vertebrates: beaver, river otter, Harter's water snake, great blue heron, snowy egret, black-crowned night-heron, white-faced ibis, kingfishers, softshell turtles, American alligator, plainbelly water snake

Major benthic insects: caddisflies (*Cheumatopsyche*), mayflies (*Choroterpes*, *Tricorythodes*), true flies (*Tanytarsus*, *Rheotanytarsus*)

Nonnative species: ≥21 fishes (grass carp, striped bass, rainbow trout, redbreast sunfish, rudd), saltcedar, Asiatic clam, nutria, water hyacinth, hydrilla, giant salvinia

Major riparian plants: baccharis, cottonwood, willow, elm, hackberry, pecan, arrowhead, eelgrass, water primrose

Special features: Llano Estacado; Blanco Canyon; dinosaur tracks in Paluxy River bank; class III rapids at Tonkawa falls west of Waco

Fragmentation: highly fragmented river, with 132 large dams in basin

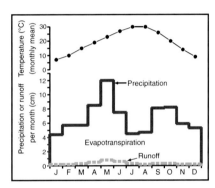

Brazos River near College Station, Texas (photo by Tim Palmer).

Sabine River

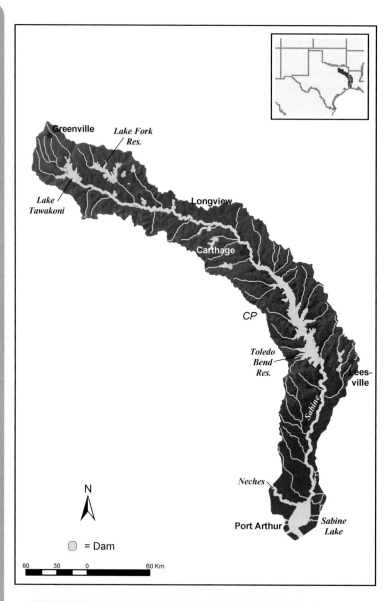

N

⬡ = Dam

60   30   0   60 Km

Relief: 198 m
Basin area: 25,268 km^2
Mean discharge: 238 m^3/s
Mean annual precipitation: 127 cm

Mean air temperature: 18°C
Mean water temperature: 21°C
No. of fish species: >104
(>88 native)
No. of endangered species: >6

Physiographic province: Coastal Plain (CP)

Major fishes: bowfin, spotted gar, alligator gar, freshwater drum, river carpsucker, blacktail redhorse, flier, creek chubsucker, grass pickerel, golden topminnow, blackspotted topminnow, brook silverside, white bass, yellow bass, dollar sunfish, redear sunfish, largemouth bass, spotted bass, black crappie, scaly sand darter, goldstripe darter, cypress darter, bigscale logperch

Major other aquatic vertebrates: alligator snapping turtle, Sabine map turtle, smooth softshell turtle, spiny softshell turtle, American alligator, Gulf Coast waterdog, pickerel frog, green water snake, western cottonmouth, wood duck, canvasback, bald eagle, white-faced ibis, anhinga, great blue heron, snowy egret, purple gallinule, belted kingfisher, river otter, nutria

Major benthic insects: caddisflies (*Hydropsyche*, *Cheumatopsyche*), true flies (*Tanytarsus*, *Rheotanytarsus*, *Stenochironomous*), mayflies (*Caenis*)

Nonnative species: ≥15 fishes (walleye, smallmouth bass, common carp, grass carp, rudd, striped bass, Mexican tetra), Asiatic clam, nutria, water hyacinth, hydrilla, giant salvinia, water spangles, Brazilian waterweed, Eurasian watermilfoil, parrotfeather, duck lettuce, alligatorweed, torpedo grass

Major riparian plants: bald cypress, sweetgum, water oak, black gum, water tupelo, magnolia, elm, cottonwood, hickory, walnut, maple, American beech, ash, palmetto, arrowhead, smartweed, buttonbush, spiderlily

Special features: 32 mussel species, scenic reach for 80 km below Toledo Bend Reservoir, Blue Elbow Swamp near Orange

Fragmentation: highly fragmented due to both low- and medium-head dams; Toledo Bend Reservoir is largest in the south

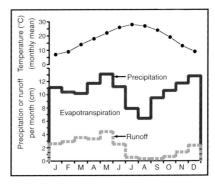

Sabine River at the Texas and Louisiana border (photo by E. Martyn).

Pecos River

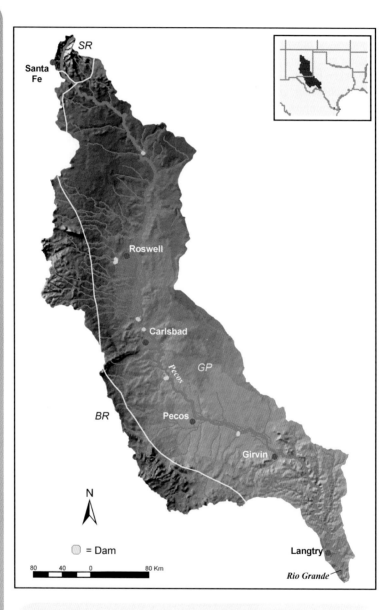

Relief: 4012 m
Basin area: 113,960 km^2
Mean discharge: 2 m^3/s (virgin ≥9 m^3/s)
Mean precipitation: 28 cm

Mean air temperature: 18°C
Mean water temperature: 21°C
No. of fish species: ≥70
No. of endangered species: >16

Physiographic provinces: Southern Rocky Mountains (SR), Basin and Range (BR), Great Plains (GP)

Major fishes: red shiner, inland silverside, Pecos pupfish, western mosquitofish, rainwater killifish, roundnose minnow, proserpine shiner, channel catfish, Rio Grande cichlid, Mexican tetra, green sunfish, largemouth bass

Major other aquatic vertebrates: great blue heron, snowy egret, black-crowned night-heron, white-faced ibis, belted kingfisher, green kingfisher, plainbelly water snake, beaver, muskrat

Major benthic insects: caddisflies (*Ithytrichia, Cheumatopsyche, Hydroptila*), mayflies (*Choroterpes, Thraulodes, Tricorythodes, Traverella*), true flies (*Tanytarsus, Dicrotendipes, Pseudochironomus, Microtendipes, Cricotopus*)

Nonnative species: ≥19 fishes (grass carp, goldfish, common carp, rudd, rainbow trout, white crappie, walleye, smallmouth bass, redear sunfish), saltcedar, Russian olive, water hyacinth, hydrilla, Eurasian watermilfoil, alligatorweed

Major riparian plants: saltcedar, mesquite, cottonwood, four-winged saltbush, Russian olive, willow

Special features: Bitter Lake National Wildlife Refuge with high biodiversity of

dragonflies and damselflies; river passes through deep gorges within limestone terrain

Fragmentation: several impoundments, including Red Bluff Reservoir upstream of Pecos, Texas; major water diversions near Pecos and Grandfalls

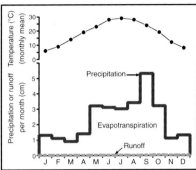

Upper Pecos River, New Mexico (photo by Tim Palmer).

Nueces River

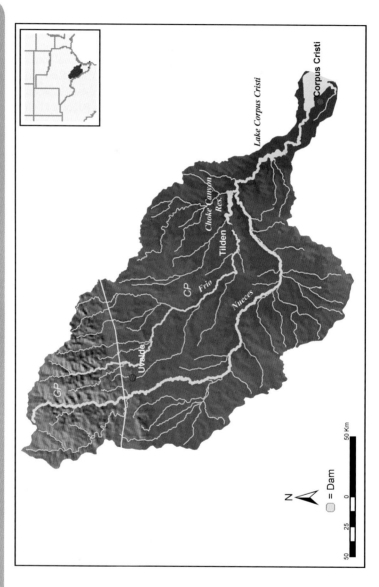

Relief: 730 m
Basin area: 43,512 km²
Mean discharge: 20 m³/s
Mean precipitation: 61 cm

Mean air temperature: 21°C
Mean water temperature: 23°C
No. of fish species: ≥66
No. of endangered species: >3

Physiographic provinces: Great Plains (GP), Coastal Plain (CP)

Major fishes: longnose gar, spotted gar, Mexican tetra, Nueces roundnose minnow, plateau shiner, Texas shiner, channel catfish, sailfin molly, western mosquitofish, largemouth bass, longear sunfish, bluegill

Major other aquatic vertebrates: beaver, great blue heron, snowy egret, white-faced ibis, belted kingfisher

Major benthic insects: caddisflies (*Chimarra, Cheumatopsyche, Hydroptila*), mayflies (*Fallceon, Dactylobaetis, Tricorythodes, Choroterpes, Thraulodes*), true flies (*Cricotopus, Rheotanytarsus, Hemerodromia, Simulium*)

Nonnative species: ≥20 fishes (goldfish, grass carp, common carp, golden shiner, rudd, rainbow trout, inland silverside, striped bass,

smallmouth bass, Guadalupe bass, walleye, Rio Grande cichlid)

Major riparian plants: pecan, Texas sugarberry, bald cypress, cottonwood, black willow, cedar elm, Texas persimmon, red mulberry, greenbrier, box elder, black walnut, American elm

Special features: arises from springs in Edwards Plateau, tricanyon area highly scenic

Fragmentation: 2 major reservoirs

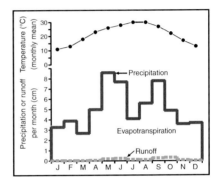

Nueces River near its headwaters at the north Barksdale Highway 335 (photo by R. Edwards).

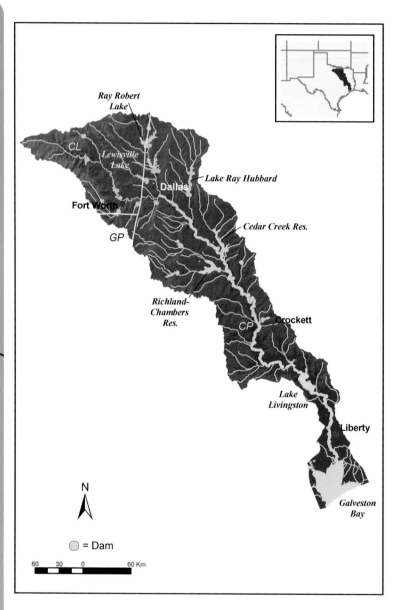

Trinity River

Relief: 362 m
Basin area: 46,540 km²
Mean discharge: 222 m³/s
Mean annual precipitation: 115 cm

Mean air temperature: 19°C
Mean water temperature: 21°C
No. of fish species: ≥99
No. of endangered species: 7

Physiographic provinces: Central Lowland (CL), Great Plains (GP), Coastal Plain (CP)

Major fishes: spotted gar, threadfin shad, channel catfish, freckled madtom, pirate perch, river carpsucker, smallmouth buffalo, red shiner, ribbon shiner, weed shiner, bullhead minnow, pugnose minnow, golden topminnow, blackstripe topminnow, western mosquitofish, dollar sunfish, largemouth bass, spotted bass, white crappie, bluntnose darter, slough darter, dusky darter

Major other aquatic vertebrates: beaver, nutria, American alligator, cottonmouth, yellow-bellied water snake, diamondback water snake, red-eared slider, Texas river cooter, common snapping turtle, cricket frog, green treefrog, gray treefrog, southern leopard frog, bald eagle, osprey, anhinga, wood duck, pintail, canvasback, great blue heron, greenback heron

Major benthic insects: caddisflies (*Cheumatopsyche*, *Chimarra*), mayflies (*Caenis*, *Isonychia*, *Heptagenia*), true flies (*Stictochironomus*)

Nonnative species: 20 fishes (grass carp, common carp, rudd, northern pike, rainbow trout, inland silverside, striped bass, redbreast sunfish, smallmouth bass, walleye, blue tilapia), nutria, Asian clam, hydrilla, water hyacinth, giant salvinia

Major riparian plants: red maple, river birch, water hickory, pecan, black hickory, hackberry, honey locust, water elm, water oak, sumac, black willow, American elm, water willow, water-pennywort, smartweed, bulrush, cattail, sedge, spikerush

Special features: Trinity River National Wildlife Refuge (permanently flooded swamps); Richland Creek and Keechi Creek Wildlife Management Areas (numerous periodically flooded oxbows)

Fragmentation: highly fragmented, with 21 major reservoirs in basin

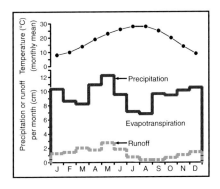

Trinity River below Livingston Dam, Texas (photo by Tim Palmer).

Neches River

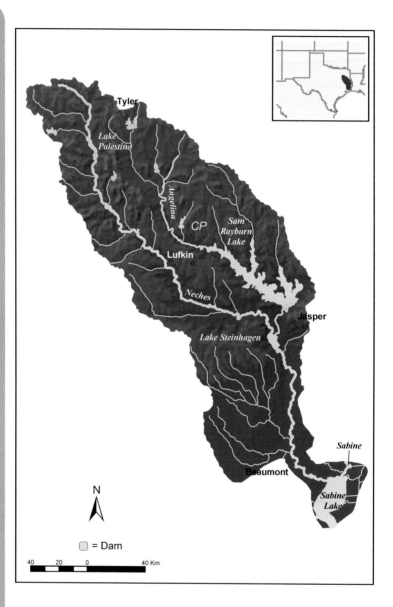

Tyler

*Lake
Palestine*

Angelina

CP

*Sam
Rayburn
Lake*

Lufkin

Neches

Jasper

Lake Steinhagen

Sabine

Beaumont

*Sabine
Lake*

N

= Dam

40   20   0   40 Km

Relief: 153 m
Basin area: 25,929 km^2
Mean discharge: 179 m^3/s
Mean precipitation: 136 cm

Mean air temperature: 19°C
Mean water temperature: 21°C
No. of fish species: ≥96
No. of endangered species: >5

Physiographic province: Coastal Plain (CP)

Major fishes: alligator gar, channel catfish, freshwater drum, blacktail redhorse, longear sunfish, spotted sunfish, pirate perch, banded pygmy sunfish, flier, spotted bass, grass pickerel, blackspot shiner, ribbon shiner, Sabine shiner, weed shiner, bullhead minnow, blackspotted topminnow, western mosquitofish, brook silverside, cypress darter, scaly sand darter, bluntnose darter

Major other aquatic vertebrates: beaver, river otter, mink, nutria, muskrat, cottonmouth, yellow-belly water snake, diamondback water snake, Graham's crayfish snake, Missouri river cooter, Sabine map turtle, alligator snapping turtle, spiny softshell turtle, green frog, American alligator, bald eagle, pied-billed grebe, anhinga, wood duck, belted kingfisher, great egret

Major benthic insects: mayflies (*Caenis*), caddisflies (*Hydropsyche*, *Cheumatopsyche*, *Hydroptila*, *Chimarra*)

Nonnative species: 15 fishes (threadfin shad, goldfish, grass carp, common carp, rudd, Mexican tetra, rainbow trout, sailfin molly, inland silverside, white bass, striped bass, redbreast sunfish, smallmouth bass, white crappie, walleye), white heelsplitter, flat floater

Major riparian plants: palmetto, bald cypress, black willow, river birch, sycamore, sweetgum, black gum, willow oak, water oak, swamp tupelo, water hickory, southern magnolia, water willow, cedar elm, cottonwood, pecan, black oak, arrowhead, water smartweed, buttonbush

Special features: Big Thicket Reserve (Man and the Biosphere Program)— remnant of large area developed on ancient sand dunes and beaches of a fossil sea; number-one scenic river in east Texas (National Park Service 1995)

Fragmentation: 3 major reservoirs in basin (Palestine, Sam Rayburn, and Steinhagen)

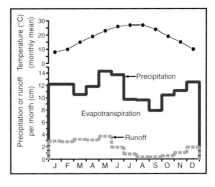

Neches River above Rt. 59, Texas (photo by Tim Palmer).

Chapter 6

Lower Mississippi River and Its Tributaries

Arthur V. Brown, Kristine B. Brown, Donald C. Jackson, and
W. Kevin Pierson

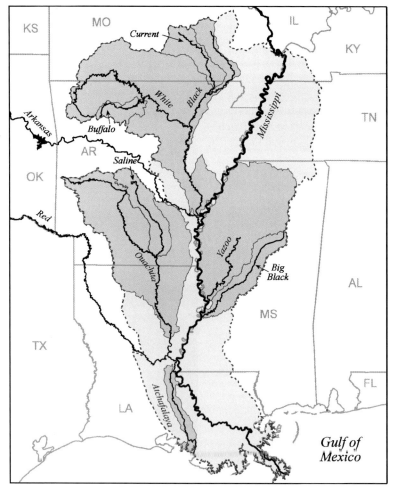

The Mississippi River basin is the largest in North America (*ca.* 3.27 million km^2: and the third largest in the world.[1] It extends from 37°N to 29°N latitude, covers nearly 14% of the continent, and drains about 40% of the conterminous United States. The Mississippi River (including its tributary, the Missouri River) at 6693 km, is second only to the Nile in length. Its estimated virgin discharge of 18,400 m^3/s ranks ninth in the world.[1] The Lower Mississippi River (LMR) extends from the confluence with the Ohio River at Cairo, Illinois, to the Gulf of Mexico (see map). Unlike other rivers considered in this book, the LMR receives most of its flow from major upstream tributaries that are considered in other chapters: the Missouri River (Chapter 10), the Upper Mississippi River (Chapter 8), and Ohio River (Chapter 9). The focal area for this chapter (460,000 km^2) excludes the upper and middle reaches of the Arkansas and Red rivers which enter this area from the west (see Chapter 7). Thus, this area includes drainages that cover most of Arkansas, Mississippi, and Louisiana, and portions of southern Missouri, western Tennessee, and western Kentucky.

The LMR has a long history of human habitation, as long as 16,000 years ago as indicated by artifacts in the basin. The last major cultural development by Native Americans initiated in the eastern United States was the Mississippian. This sedentary society, which built earthen mounds with wooden buildings on them, developed along the LMR during the 8th through the middle of the 13th century. At the time Europeans arrived, clans of Choctaw, Chippewa, Koroa, Taensa, Chickasaw, and many others lived in the LMR valley. Although first explored by the Spanish beginning with Hernando De Soto, the LMR was first settled by the French. La Salle claimed the river basin for France in 1682. France ceded it to Spain in 1762, but in 1800, Napoleon Bonaparte managed to take it back again. France sold the "Louisiana Purchase" to the United States in 1803, doubling its size.

In this chapter, we describe the main stem LMR as well as several of its tributaries, exclusive of the Red and Arkansas rivers. The largest of these tributaries are the White, Ouachita, and Yazoo, each with a basin area of at least 35,000 km^2 and a mean discharge of >500 m^3/s. The Atchafalaya is actually a distributary of the LMR that typically carries about 25% of the LMR flow to the Gulf of Mexico. Also described are several smaller tributaries: the Buffalo, Big Black, Saline, and Current.

---

1 Leopold, L. M. 1994. A view of the river. Harvard University Press, Cambridge, Massachusetts.

Lower Mississippi River

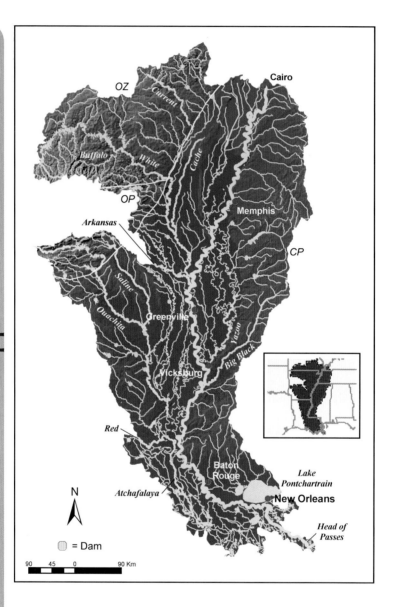

Relief: 826 m (LMR only), 4141 m (entire basin)

Basin area: $3.27 \times 10^6\,\mathrm{km}^2$

Mean discharge: $18{,}400\,\mathrm{m}^3/\mathrm{s}$

Mean annual precipitation: 140 cm (LMR only), 94 cm (entire basin)

Mean air temperature: 17°C (LMR only)

Mean water temperature: 16°C

No. of fish species: 375 (entire Mississippi basin)

No. of endangered species: 6

Physiographic provinces: Coastal Plain (CP), Ouachita Province (OP), Ozark Plateaus (OZ)

Major fishes: longnose gar, shortnose gar, shovelnose sturgeon, bowfin, gizzard shad, threadfin shad, central silvery minnow, speckled chub, silver chub, emerald shiner, river shiner, silverband shiner, mimic shiner, river carpsucker, blue sucker, smallmouth buffalo, blue catfish, channel catfish, flathead catfish, inland silverside, white bass, sauger, freshwater drum

Major other aquatic vertebrates: American alligator, snapping turtles, softshell turtles, Mississippi mud turtle, red-eared turtle, bullfrog, pigfrog, southern leopard frog, beaver, muskrat, river otter, nutria, great blue heron

Major benthic insects: mayflies (*Pentagenia, Tortopus, Stenonema, Baetis*), caddisflies (*Hydropsyche, Potamyia*), true flies (*Chaoborus, Rheotanytarsus*)

Nonnative species: Asian clam, zebra mussel, common carp, grass carp, silver carp, bighead carp, striped bass, greenhouse frog, nutria, alligatorweed, wild taro, water hyacinth, Peruvian water grass, Eurasian watermilfoil

Major riparian plants: willow, bald cypress, water tupelo, Nuttall's oak, swamp chestnut oak, overcup oak, sweetgum, ash, river birch, cottonwood, Eastern hophornbeam, water hickory, pecan, buttonbush, drummond maple, palmettos, maidencane, alligatorweed

Special features: third largest river basin in world

Fragmentation: no dams on LMR main stem but highly modified

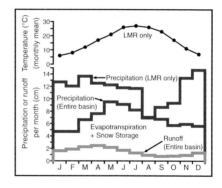

Mississippi River at New Orleans, Louisiana (photo by Tim Palmer).

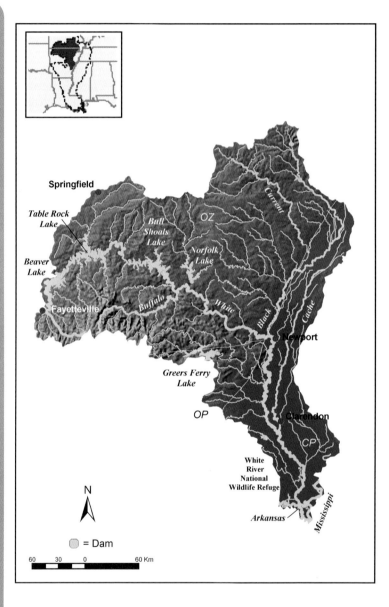

Springfield

Table Rock
Lake

Bull
Shoals
Lake

OZ

Current

Norfolk
Lake

Beaver
Lake

Fayetteville

Buffalo

White

Black

Cache

Newport

Greers Ferry
Lake

OP

Clarendon

CP

White
River
National
Wildlife Refuge

N

Arkansas

Mississippi

⬤ = Dam

60   30   0   60 Km

White River

Relief: 731 m
Basin area: 72,189 km²
Mean discharge: 979 m³/s
Mean annual precipitation: 117 cm

Mean air temperature: 15°C
Mean water temperature: 19°C
No. of fish species: 163
No. of endangered species: 15

Physiographic provinces: Ozark Plateaus (OZ), Ouachita Province (OP), Coastal Plain (CP)
Major fishes: longnose gar, gizzard shad, central stoneroller, Mississippi silvery minnow, Ozark minnow, bullhead minnow, emerald shiner, bigeye shiner, duskystripe shiner, blacktail shiner, mimic shiner, northern hogsucker, black redhorse, channel catfish, blue catfish, Ozark bass, longear sunfish, smallmouth bass, white crappie, rainbow darter, orangethroat darter, banded sculpin
Major other aquatic vertebrates: American alligator, snapping turtle, Mississippi mud turtle, map turtle, slider, softshell turtles, cottonmouth, midland water snake, yellow-bellied water snake, cricket frog, bullfrog, green frog, southern leopard frog, pickerel frog, muskrat, beaver, river otter, raccoon, mink, great blue heron, belted kingfisher
Major benthic insects: mayflies (*Tortopus, Pentagenia, Baetis, Stenonema, Caenis*), stoneflies (*Acroneuria*), caddisflies (*Chimarra, Hydropsyche, Agapetus*)
Nonnative species: Asian clam, zebra mussel, grass carp, common carp, silver carp, bighead carp, yellow perch, sauger, walleye, redeye bass, striped bass, white bass, American shad, threadfin shad, northern pike, muskellunge, chain pickerel, brown bullhead, cutthroat trout, rainbow trout, brown trout, brook trout, lake trout, Eurasian watermilfoil, water hyacinth, duck lettuce

Major riparian plants: willows, witchhazel, American sycamore, river birch, red maple, buttonbush, cottonwood, American elm, green ash, box elder, sugarberry, sweetgum, Nuttall's oak, bald cypress, water tupelo, hickories, river cane, poison ivy, greenbrier, cucumber vine, smartweed
Special features: karst topography in headwaters, extensive bottomland hardwood forest floodplains; two tributaries are National Rivers
Fragmentation: 4 large dams on main stem, 3 large dams on tributaries

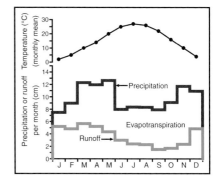

White River near Flippen, Arkansas (photo by J. Powers).

Buffalo National River

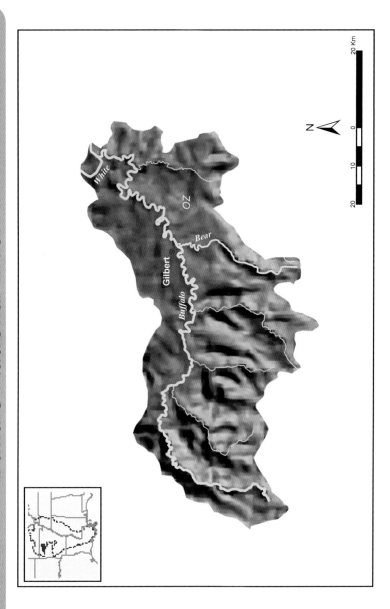

Relief: 666 m
Basin area: 3465 km²
Mean discharge: 48 m³/s
Mean annual precipitation: 107 cm

Mean air temperature: 15°C
Mean water temperature: 13°C
No. of fish species: >66
No. of endangered species: 10

Physiographic province: Ozark Plateaus (OZ)
Major fishes: longear sunfish, Ozark bass, largescale stoneroller, duskystripe shiner, rosyface shiner, telescope shiner, bigeye shiner, Ozark madtom, slender madtom, channel catfish, banded sculpin, greenside darter, rainbow darter, Arkansas saddled darter, yoke darter
Major other aquatic vertebrates: beaver, river otter, mink, common snapping turtle, map turtle, slider, midland smooth softshell turtle, western cottonmouth, midland water snake, red river waterdog, cricket frog, bullfrog, green frog
Major benthic insects: mayflies (*Pseudocloeon, Heptagenia, Stenonema, Ephemerella, Isonychia, Baetis, Ephoron*), stoneflies (*Perlesta*), caddisflies (*Agapetus, Cheumatopsyche*)
Nonnative species: freshwater jellyfish, Asian clam, common carp, fathead minnow, western mosquitofish, largemouth bass, smallmouth bass, walleye, rainbow trout
Major riparian plants: American elm, green ash, silver maple, box elder, American sycamore, river birch, black willow, Ward's willow, sandbar willow, cottonwood, sweetgum, witchhazel, buttonbush, giant river cane, sea oats, sedges, water willow
Special features: almost pristine main stem protected as National River; free-flowing
Fragmentation: none

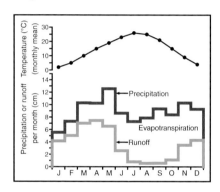

Buffalo National River, Arkansas (photo by A. C. Benke).

Big Black River

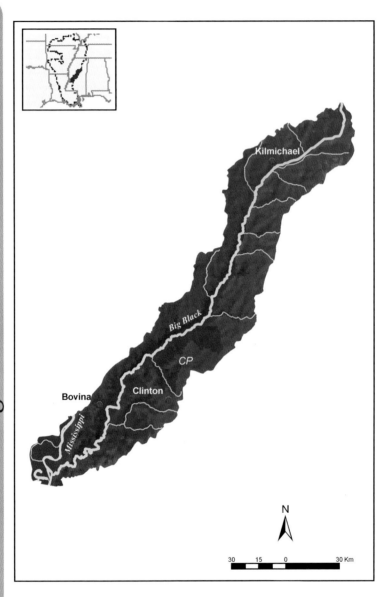

Relief: 152 m
Basin area: 8770 km²
Mean discharge: 107 m³/s
Mean annual precipitation: 135 cm

Mean air temperature: 17.7°C
Mean water temperature: 17.2°C
No. of fish species: 112 (native)
No. of endangered species: 4

Physiographic province: Coastal Plain (CP)

Major fishes: flathead catfish, blue catfish, channel catfish, smallmouth buffalo, bigmouth buffalo, black buffalo, freshwater drum, white crappie, black crappie, largemouth bass, gizzard shad, bluegill, longnose gar, spotted gar, blue sucker, paddlefish, blacktail shiner, emerald shiner, striped shiner, creek chubsucker, freckled madtom, blackspotted topminnow, central stoneroller, scaly sand darter, slough darter, logperch darter, dusky darter

Major other aquatic vertebrates: beaver, river otter, cottonmouth, red-bellied water snake, common snapping turtle, alligator snapping turtle, slider, Mississippi map turtle, great blue heron, green heron, little blue heron, wood duck, mallard, kingfisher

Major benthic insects: mayflies (*Baetis, Cinygmula, Tricorythodes*), caddisflies (*Hydropsyche, Cheumatopsyche, Nectopsyche*), true flies (*Prosimulium*)

Nonnative species: Asian clam, common carp, grass carp, goldfish, striped bass, bluespotted sunfish, dotted duckweed, parrot feather, sacred lotus, water lettuce

Major riparian plants: American sycamore, red maple, river birch, black willow, green ash, cottonwood, water oak, willow

oak, water tupelo, bald cypress, honey suckle, American lotus, waterlily, water shield, pondweed, cattail, water primrose, alligatorweed

Special features: only free-flowing river in Mississippi that flows directly into Mississippi River; intact forested floodplain; numerous Civil War relics (sunken gunboats)

Fragmentation: no dams; only portions of lowermost reaches impacted by channel modification

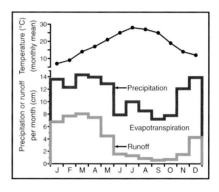

Big Black River, Mississippi (photo by D. Jackson).

Yazoo River

Relief: 195 m
Basin area: 35,000 km²
Mean discharge: 523 m³/s
Mean annual precipitation: 134 cm

Mean air temperature: 17.6°C
Mean water temperature: 21.5°C
No. of fish species: 119 (native)
No. of endangered species: 15

Physiographic province: Coastal Plain (CP)

Major fishes: flathead catfish, blue catfish, channel catfish, smallmouth buffalo, bigmouth buffalo, freshwater drum, white crappie, black crappie, largemouth bass, gizzard shad, bluegill, longnose gar, spotted gar, blue sucker, paddlefish, black buffalo, striped bass

Major other aquatic vertebrates: beaver, river otter, cottonmouth, bullfrog, common snapping turtle, alligator snapping turtle, Mississippi map turtle, slider, great blue heron, wood duck

Major benthic insects: mayflies (*Baetis*, *Caenis*, *Cinygmula*, *Stenonema*, *Potamanthus*), caddisflies (*Cheumatopsyche*, *Hydropsyche*, *Nectopsyche*), true flies (*Chaoborus*, *Simulium*, *Prosimulium*, *Hemerodromia*)

Nonnative species: freshwater jellyfish, Asian clam, common carp, grass carp, bighead carp, goldfish, white bass, striped bass, yellow perch, walleye, fathead minnow, American shad, tilapia, alligatorweed, water hyacinth, dotted duckweed, parrot feather, water lettuce

Major riparian plants: American sycamore, red maple, river birch, black willow, green ash, cottonwood, water oak, willow oak, water tupelo, bald cypress, honey suckle, water primrose, spikerush, alligatorweed, water shield, cattail, American lotus, waterlily, pondweed

Special features: tremendously productive recreational and "artisanal" fisheries for catfish; floodplains internationally important overwintering areas for migratory waterfowl

Fragmentation: major dams on tributaries (Coldwater, Little Tallahatchie, Yocona, Yalobusha–Skuna); extensive channelization

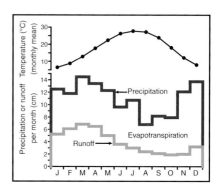

Yazoo River, Mississippi (photo by D. Jackson).

Atchafalaya River

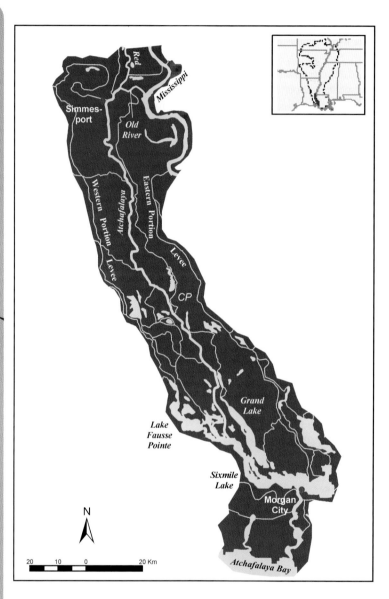

Relief: 15 m
Basin area: 8345 km^2
Mean discharge: 5178 m^3/s
Mean annual precipitation: 153 cm

Mean air temperature: 20°C
Mean water temperature: 22°C
No. of fish species: 181
No. of endangered species: 9

Physiographic province: Coastal Plain (CP)

Major fishes: redear sunfish, bluegill, smallmouth buffalo, white crappie, black crappie, largemouth bass, warmouth, white bass, spotted gar, alligator gar, channel catfish, blue catfish, gizzard shad, threadfin shad, freshwater drum, bowfin, several shiners, mosquitofish, inland silverside

Major other aquatic vertebrates: American alligator, western cottonmouth, yellow-bellied water snake, bullfrog, pigfrog, southern leopard frog, beaver, muskrat, river otter, mink, nutria, great blue heron, green heron, cormorant, belted kingfisher

Major benthic insects: mayflies (*Tortopus*, *Pentagenia*), caddisflies (*Hydropsyche*), true flies (*Coelotanypus*, *Polypedilum*, *Chaoborus*)

Nonnative species: Asian clam, zebra mussel, grass carp, silver carp, bighead carp, common carp, nutria, water hyacinth, Eurasian watermilfoil, hydrilla, alligatorweed, wild taro, horsefly's eye, Brazilian water weed, dotted duck weed, marshweed, Uruguay seedbox, parrot feather, brittle naiad

Major riparian plants: bald cypress, black willow, water tupelo, drummond maple, cottonwood, river birch, American sycamore, sweetgum, sugarberry, buttonbush, ash, pondweed, frogbite, bladderwort, maidencane, cattails, dwarf spikerush, purple ammania, palmetto

Special features: 3rd largest continuous wetland area in United States

Fragmentation: no dams, but extensive levees; discharge controlled by dams and floodgates linking Atchafalaya to Mississippi and Red rivers

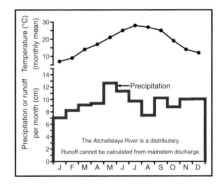

Delta of the Atchafalaya River on the Gulf of Mexico (photo by A. Belala, USACE).

Ouachita River

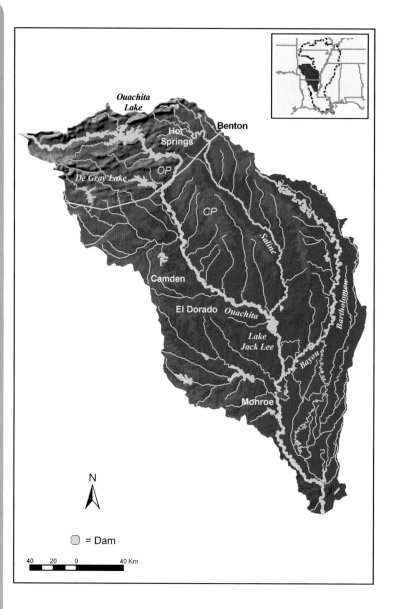

N

⬤ = Dam

40   20   0   40 Km

Relief: 810 m
Basin area: 64,454 km²
Mean discharge: 843 m³/s
Mean annual precipitation: 130 cm

Mean air temperature: 17°C
Mean water temperature: 16°C
No. of fish species: 80
No. of endangered species: 12

Physiographic provinces: Ouachita Province (OP), Coastal Plain (CP)

Major fishes: spotted gar, longnose gar, northern hog sucker, spotted sucker, black redhorse, golden redhorse, central stoneroller, bigeye shiner, rosyface shiner, redfin shiner, steelcolor shiner, bluntnose minnow, blackspotted topminnow, channel catfish, freckled madtom, brook silverside, green sunfish, spotted bass, orangebelly darter, greenside darter, channel darter

Major other aquatic vertebrates: snapping turtle, stinkpot turtle, Mississippi mud turtle, Ouachita map turtle, southern painted turtle, midland smooth softshell turtle, western spiny softshell turtle, western cottonmouth, broad-banded water snake, yellow-bellied water snake, diamondback water snake, green water snake, bullfrog, green frog, southern leopard frog, great blue heron, belted kingfisher, river otter, mink

Major benthic insects: stoneflies (*Amphinemura*), mayflies (*Isonychia, Caenis, Stenonema*), caddisflies (*Hydroptila, Agapetus, Chimarra*)

Nonnative species: Asian clam, threadfin shad, grass carp, common carp, silver carp, bighead carp, northern pike, muskellunge, chain pickerel, blue catfish, white catfish, striped bass, sauger, walleye, rainbow trout, brown trout, brook trout, alligatorweed,

wild taro, water hyacinth, yellow iris, parrot feather, water lettuce, hydrilla, Eurasian watermilfoil

Major riparian plants: American sycamore, sweetgum, shortleaf pine, loblolly pine, American hornbeam, eastern hophornbeam, red oak, beech, American holly, Ozark witchhazel, water tupelo, poison ivy, greenbrier, smartweed

Special features: geothermal hot springs in basin; natural reservoir formed by Felsenthal basin

Fragmentation: 5 impoundments on main stem; 1 large impoundment on each of 2 major tributaries

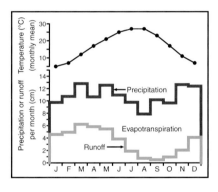

Ouachita River, Columbia Lock and Dam, Louisiana (photo by B. Emerson, USACE).

Saline River

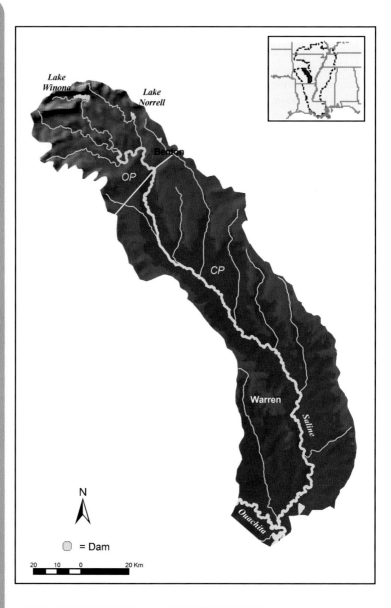

= Dam

20   10   0   20 Km

Relief: 532 m   Mean air temperature: 17°C
Basin area: 5465 km²   Mean water temperature: 17°C
Mean discharge: 89 m³/s   No. of fish species: 85
Mean annual precipitation: 130 cm   No. of endangered species: 4

Physiographic provinces: Ouachita Province (OP), Coastal Plain (CP)

Major fishes: smallmouth bass, largemouth bass, spotted bass, warmouth, shadow bass, longear sunfish, bluegill, green sunfish, banded pygmy sunfish, black crappie, channel catfish, Ouachita madtom, cypress darter, taillight shiner, peppered shiner, redfin shiner, big eye shiner, striped shiner, steelcolor shiner

Major other aquatic vertebrates: snapping turtle, Ouachita map turtle, midland smooth softshell turtle, midland water snake, yellow-bellied water snake, diamondback water snake, green water snake, western cottonmouth, red river waterdog, green frog, bullfrog, southern leopard frog, pickerel frog, belted kingfisher, great blue heron, muskrat, beaver, mink, river otter

Major benthic insects: mayflies (*Stenonema*, *Isonychia*), stoneflies (*Neoperla*, *Amphinemura*), caddisflies (*Chimarra*, *Cheumatopsyche*)

Nonnative species: freshwater jellyfish, Asian clam, grass carp, common carp, goldfish, walleye, chain pickerel, blue catfish, threadfin shad, fathead minnow

Major riparian plants: water oak, willow oak, red oak, sweetgum, American sycamore, black willow, river birch, buttonbush, smooth alder, eastern hophornbeam, American hornbeam, common winterberry, haws, water willow, smartweed

Special features: one of the few natural free-flowing rivers in conterminous 48 states (last in Ouachita Mountain area); excellent float and fishing river

Fragmentation: no major dams, 2 small ones in headwaters

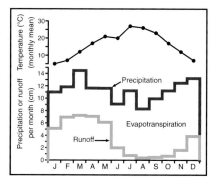

Saline River near Malvern, Arkansas (photo by D. Jackson).

Current River

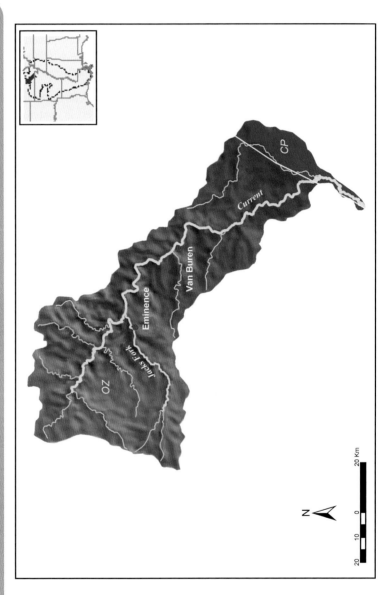

Relief: 372 m
Basin area: 6776 km²
Mean discharge: 77 m³/s
Mean annual precipitation: 123 cm

Mean air temperature: 15°C
Mean water temperature: 17°C
No. of fish species: 117
No. of endangered species: 9

Physiographic provinces: Ozark Plateaus (OZ), Coastal Plain (CP)
Major fishes: smallmouth bass, rock bass, longear sunfish, northern hog sucker, central stoneroller, rosyface shiner, telescope shiner, Arkansas saddled darter, greenside darter, rainbow darter, fantail darter, stargazing darter, mountain madtom, blackspotted topminnow
Major other aquatic vertebrates: snapping turtle, softshell turtle, Mississippi map turtle, midland water snake, western cottonmouth, green frog, beaver, river otter, muskrat, raccoon, mink, great blue heron, belted kingfisher, green-backed heron
Major benthic insects: mayflies (*Stenonema*, *Baetis*), stoneflies (*Neoperla*, *Leuctra*), caddisflies (*Hydropsyche*, *Ceratopsyche*)
Nonnative species: Asian clam, common carp, chain pickerel, grass pickerel, white bass, walleye, rainbow trout, brown trout
Major riparian plants: American sycamore, box elder, American elm, winged elm, slippery elm, black willow, river birch, hackberry, silver maple, sugar maple, bur oak, green ash, white ash, common witchhazel, spicebush, pawpaw, American hornbeam, flowering dogwood, hawthorns, poison ivy, water willow, smartweed
Special features: 161 km protected as Ozark

National Scenic Riverways; numerous large springs provide 60% of base flow; one of few natural free-flowing rivers in conterminous 48 states
Fragmentation: none

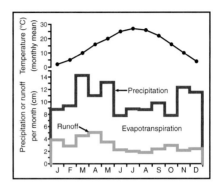

Current River above Van Buren, Missouri (photo by Tim Palmer).

Chapter 7

Southern Plains Rivers

William J. Matthews, Caryn C. Vaughn, Keith B. Gido,
and Edie Marsh-Matthews

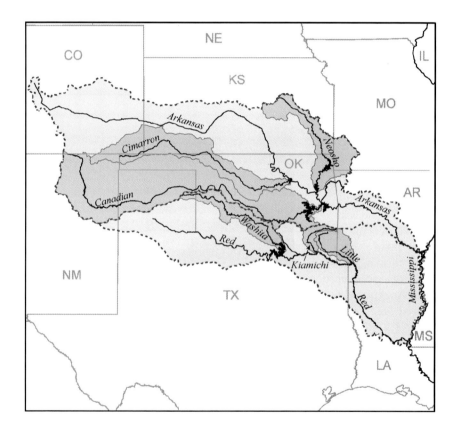

Two large, separate river basins, the Arkansas and the Red, drain the southern Great Plains region of the United States south of the Kansas River and north of the Texas-Gulf coastal drainages (see map). All major rivers in the region drain generally from northwest to southeast, and are tributaries of the Mississippi River. The Southern Plains region includes all of Oklahoma, much of western and central Arkansas, and parts of eastern New Mexico, Colorado, Kansas, north Texas, and western and central Louisiana. The region is characterized by shortgrass prairie in the west, mixed or tallgrass prairie in the midsection, and forests in the east. The largest rivers (Arkansas, Canadian, Red, Washita, Cimarron) have in common that upper main stems lack flow at times, and that mid and downstream reaches are wide, shallow, and sand or mud-bottomed. They are some of the hottest and harshest aquatic habitats on Earth, with water temperatures reaching near 40°C when exposed to full sun under low flow conditions.

People were occupying the Arkansas and Red basins by 11,500 to 10,000 years ago, as hunters of the last ice-age large animals. The Ouachita Mountains and Ozarks parts of the Arkansas and Red basins have a long history of occupation by hunter-gatherers, as early as 10,500 years ago. By 900 years ago major populations occupied the fertile valleys along several rivers in eastern Oklahoma and adjacent Arkansas. To the west, bison hunters and farmers were found along several tributaries of the plains. The Arkansas and Red basins first came under European control under the claims of Spanish explorers like Coronado and De Soto in the 1500s. By the early 1700s many French explorers, trappers, and traders came into the region, making contact with, and in many cases, marrying native people. Spain was recognized as owner of the region in treaties of 1762–1763, but transferred ownership of "Louisiana" to France in negotiations in 1800–1802. The United States purchased "Louisiana" in 1803, by which most of the Red and Arkansas river basins, along with the Missouri and western Mississippi basins, became permanently owned by the United States.

In addition to the Arkansas and Red rivers, this chapter describes several of their tributaries. Both the Arkansas and Red rivers drain very large basins ($415{,}000\,km^2$ and $170{,}000\,km^2$, respectively), and have high mean discharges (1004 and $852\,m^3/s$). The tributaries include two large rivers of the Great Plains (Canadian, Cimarron) having relatively low discharge for their size and two relatively large rivers draining the Central Lowlands and Ozark Plateaus (Little, Neosho) with discharge approaching or exceeding $200\,m^3/s$. Two smaller rivers, the Washita and Kiamichi, also are described.

Arkansas River

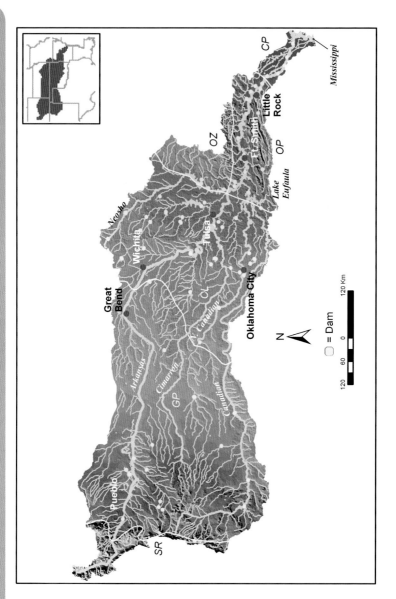

Relief: 4340 m
Basin area: 414,910 km²
Mean discharge: 1004 m³/s
Mean annual precipitation:
71.8 cm

Mean air temperature: 15°C
Mean water temperature: 17.9°C
No. of fish species: 171 (141 native)
No. of endangered species: 2

Physiographic provinces: Southern Rocky Mountains (SR), Great Plains (GP), Central Lowland (CL), Ozark Plateaus (OZ), Ouachita Province (OP), Coastal Plain (CP)

Major fishes: paddlefish, gars, gizzard shad, red shiner, river shiner, emerald shiner, plains minnow, smallmouth buffalo, bigmouth buffalo, river carpsucker, channel catfish, flathead catfish, plains killifish, western mosquitofish, white bass, largemouth bass, spotted bass, sunfishes, river darter

Major other aquatic vertebrates: plains leopard frog, American bullfrog, snapping turtle, spiny softshell turtle, smooth softshell turtle, common slider, painted turtle, northern water snake, diamondback water snake

Major benthic insects: mayflies (*Caenis, Hexagenia, Stenonema*), caddisflies (*Cheumatopsyche, Hydropsyche*), true flies (*Polypedilum, Glyptotendipes*)

Nonnative species: Asian clam, zebra mussel, ~30 fish species (common carp, grass carp, striped bass), nutria

Major riparian plants: silver maple, box elder, hackberry, cottonwood, willow, cattails, American bulrush

Special features: arises as mountain river, almost disappears in western Kansas due to water withdrawal

Fragmentation: 5 major reservoirs on main stem, 17 locks and dams; part of Kerr-McClellan Navigation System

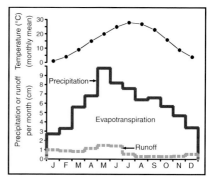

Arkansas River from Mt. Petit Jean, Arkansas (photo by M. Keckhaver/Encyclopedia of Arkansas History & Culture).

Canadian River

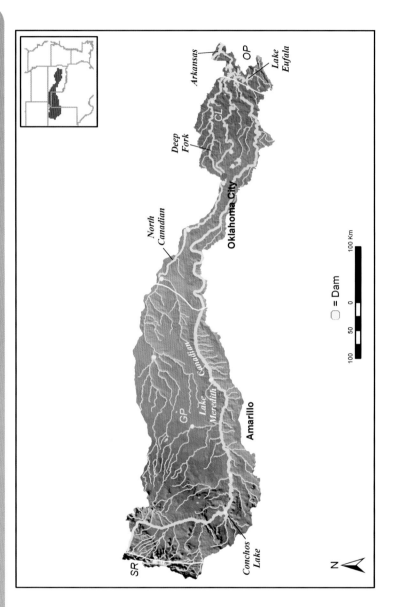

Relief: 4132 m
Basin area: 122,070 km^2
Mean discharge: 174 m^3/s
Mean annual precipitation: 53 cm

Mean air temperature: 15°C
Mean water temperature: 18°C
No. of fish species: 63 (native)
No. of endangered species: 2

Physiographic provinces: Southern Rocky Mountains (SR), Great Plains (GP), Central Lowland (SL), Ouachita Province (OP)

Major fishes: gizzard shad, red shiner, Arkansas River shiner, emerald shiner, plains minnow, bluntnose minnow, fathead minnow, plains killifish, western mosquitofish, river carpsucker, channel catfish, white bass, largemouth bass, longear sunfish, green sunfish

Major other aquatic vertebrates: snapping turtle, yellow mud turtle, stinkpot turtle, smooth softshell turtle, beaver

Major benthic insects: true flies (*Bezzia*, *Chironomus*, *Cryptochironomus*), mayflies (*Tricorythodes*, *Caenis*), caddisflies (*Cheumatopsyche*)

Nonnative species: Asian clam, Red River pupfish, inland silversides, common carp, blue tilapia, saltcedar

Major riparian plants: silver maple, box elder, American elm, hackberry, sandbar willow, ash, cottonwood, saltcedar

Special features: crosses arid grasslands in west, sometimes desiccating, mesic forest in east; shallow, shifting "sand bed" rivers create harsh environments, limiting richness and persistence of fauna

Fragmentation: 4 impoundments on main stem

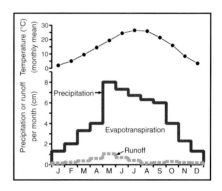

South Canadian River, Oklahoma (photo by W. J. Matthews).

Red River

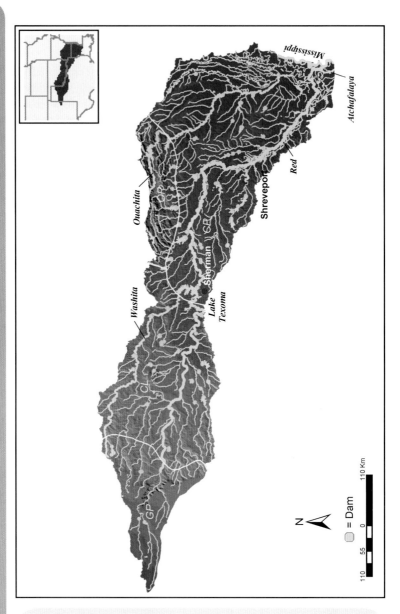

Relief: 1347 m
Basin area: 169,890 km^2
Mean discharge: 852 m^3/s
Mean annual precipitation: 82 cm

Mean air temperature: 18°C
Mean water temperature: 19.3°C
No. of fish species: 171 (152 native)
No. of endangered species: 1

Physiographic provinces: Great Plains (GP), Central Lowland (CL), Ouachita Province (OP), Coastal Plain (CP)

Major fishes: alligator gar, longnose gar, gizzard shad, red shiner, emerald shiner, Red River shiner, chub shiner, emerald shiner, blacktail shiner, bluntnose minnow, plains minnow, blue sucker, smallmouth buffalo, river carpsucker, channel catfish, blue catfish, plains killifish, Red River pupfish, sunfishes, white bass, largemouth bass, bigscale logperch

Major other aquatic vertebrates: alligator snapping turtle, common slider, spiny softshell turtle, false map turtle, yellow mud turtle, plain-bellied water snake, cottonmouth, American alligator, great blue heron, beaver, muskrat, nutria

Major benthic insects: true flies (*Glyptotendipes, Dicrotendipes, Chironomus*), mayflies (*Hexagenia, Caenis, Stenonema*), caddisflies (*Cyrnellus, Hydropsyche*)

Nonnative species: Asian clam, nutria, striped bass, walleye, threadfin shad, inland silversides, common carp, grass carp

Major riparian plants: cottonwood, willows, box elder, silver maple, slippery elm, sweetgum, post oak

Special features: spans gradient from driest to some of wettest climatic conditions in North America; high salinity in headwaters, frequently drying; Great Raft,

once a logjam of gigantic proportions upstream from Shreveport

Fragmentation: one major impoundment (Lake Texoma) on main stem; 4 locks and dams in Louisiana

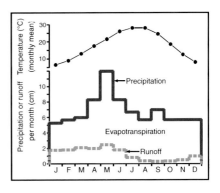

Red River, Oklahoma (photo by W. J. Matthews).

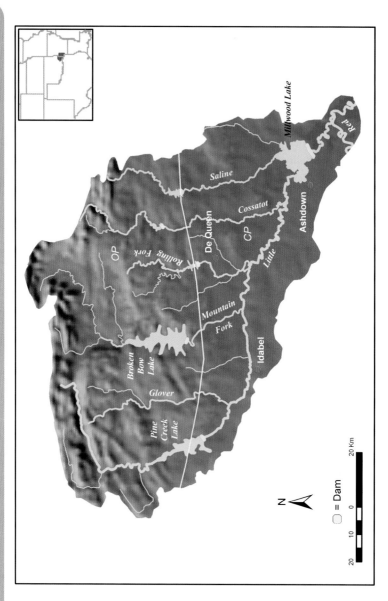

Little River

Relief: 741 m
Basin area: 10,720 km²
Mean discharge: 183 m³/s
Mean annual precipitation: 123 cm

Mean air temperature: 16°C
Mean water temperature: 16.5°C
No. of fish species: 110
No. of endangered species: 3

Physiographic provinces: Ouachita Province (OP), Coastal Plain (CP)
Major fishes: gars, rocky shiner, blacktail shiner, central stoneroller, river redhorse, golden redhorse, blackstriped topminnow, grass pickerel, flier, bantam sunfish, pirate perch, dusky darter, crystal darter, orangethroat darter, orangebelly darter, largemouth bass, spotted bass, sunfishes, leopard darter
Major other aquatic vertebrates: snapping turtle, common slider, razor-backed musk turtle, diamondback water snake, northern water snake, cottonmouth, swamp rabbit, beaver, river otter, mink
Major benthic insects: mayflies (*Stenonema*, *Ephemerella*, *Heptagenia*, *Isonychia*), stoneflies (*Acroneuria*, *Neoperla*), caddisflies (*Cheumatopsyche*, *Helicopsyche*, *Chimarra*)
Nonnative species: Asian clam, brown trout, rainbow trout, common carp; grass carp and striped bass likely

Major riparian plants: river birch, sycamore, smooth alder, sugar maple, box elder, willow oak, blue beech, bald cypress
Special features: some of last well-preserved upland rivers in central United States; regional "hot spot" of biodiversity
Fragmentation: 2 reservoirs on main stem (Pine Creek and Millwood); 4 large reservoirs on main tributaries

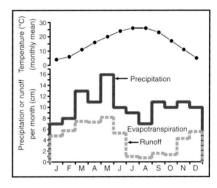

Little River near Tecumseh, Oklahoma (photo by NOAA).

Cimarron River

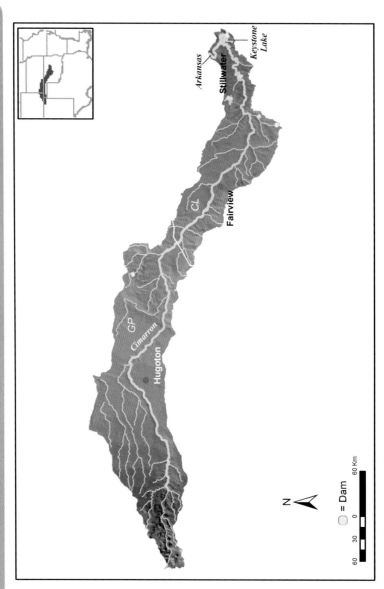

Relief: 2036 m
Basin area: 50,540 km²
Mean discharge: 42 m³/s
Mean annual precipitation: 55 cm

Mean air temperature: 15°C
Mean water temperature: 18.4°C
No. of fish species: 48
No. of endangered species: 2

Physiographic provinces: Great Plains (GP), Central Lowland (CL)

Major fishes: red shiner, plains minnow, plains killifish, gizzard shad, white bass, channel catfish, western mosquitofish; Arkansas River shiner now much reduced in abundance

Major other aquatic vertebrates: snapping turtle, beaver

Major benthic insects: mayflies (*Caenis*, *Baetis*), caddisflies (*Cheumatopsyche*, *Hydropsyche*)

Nonnative species: Asian clam, Red River shiner, striped bass, saltcedar

Major riparian plants: silver maple, box elder, ash, hackberry, cottonwood, sandbar willow, black willow, saltcedar, American elm

Special features: drains some of most arid lands of southern Great Plains;

long reaches of western main stem intermittent; harsh conditions, but relatively diverse fish fauna

Fragmentation: 2 large reservoirs on main stem; other fragmentation by natural or human-enhanced desiccation of main-stem reaches

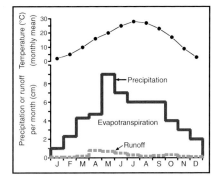

Cimarron River from Highway 77 (photo by B. Caldwell).

Neosho (Grand) River

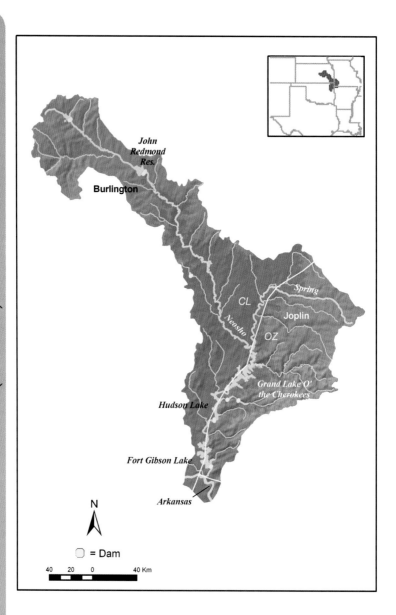

Relief: 325 m
Basin area: 54,550 km^2
Mean discharge: 254 m^3/s
Mean annual precipitation: 91 cm

Mean air temperature: 12°C
Mean water temperature: 15.4°C
No. of fish species: 94 native
No. of endangered species: 3

Physiographic provinces: Central Lowland (CL), Ozark Plateaus (OZ)

Major fishes: upstream: Topeka shiner, orangethroat darter, cardinal shiner, southern redbelly dace, endemic Neosho madtom; downstream: paddlefish, gizzard shad, numerous native minnows, smallmouth buffalo, river carpsucker, white bass, largemouth bass, sunfishes

Major other aquatic vertebrates: hellbender (threatened in Kansas), mudpuppy, snapping turtle, spiny softshell turtle, smooth softshell turtle, common slider, false map turtle, Ouachita map turtle, painted turtle, diamondback water snake

Major benthic insects: caddisflies (*Hydropsyche*, *Potamyia*), true flies (*Glyptotendipes*)

Nonnative species: Asian clam, common carp, rainbow trout

Major riparian plants: silver maple, box elder, red maple, river birch, hackberry, pecan, eastern swamp privet, ash, blackgum, sycamore, cottonwood, pin oak, American elm

Special features: drains unique uplifted region of Kansas known as "Flint Hills"; streams comprising clear water "outposts" disjunct from and containing species common to Ozark Plateaus

Fragmentation: 4 impoundments on main stem

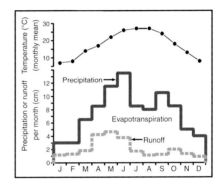

Neosho River near Neosho Rapids, Kansas (photo by NOAA).

Washita River

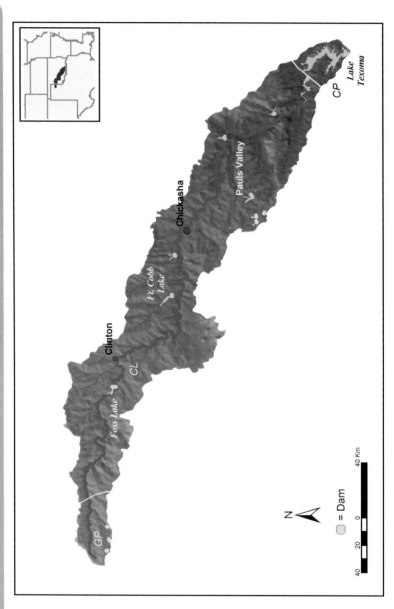

Relief: 714 m
Basin area: 20,230 km²
Mean discharge: 44 m³/s
Mean annual precipitation: 76 cm

Mean air temperature: 16°C
Mean water temperature: 18.4°C
No. of fish species: 51
No. of endangered species: 0

Physiographic provinces: Great Plains (GP), Central Lowland (CL), Coastal Plain (CP)

Major fishes: gizzard shad, speckled chub, channel catfish, longear sunfish, green sunfish, bluegill, red shiner; carpsuckers common in lower river

Major other aquatic vertebrates: common slider, false map turtle, plain-bellied water snake, beaver

Major benthic insects: mayflies (*Baetis*, *Choroterpes*), caddisflies (*Hydropsyche*, *Hydroptila*), true flies (chironomid midges)

Nonnative species: Asian clam, striped bass, threadfin shad, inland silversides, common carp, saltcedar

Major riparian plants: silver maple, box elder, ash, hackberry, cottonwood, bur oak, sandbar willow, black willow, saltcedar, American elm

Special features: lower main stem one of most turbid rivers in North America, extremely heavy load of silt or clay; very muddy bottoms; large snag piles common, likely to be major habitat

Fragmentation: 2 impoundments on main stem

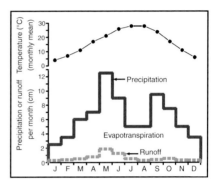

Washita River, Oklahoma (photo by W. J. Matthews).

Kiamichi River

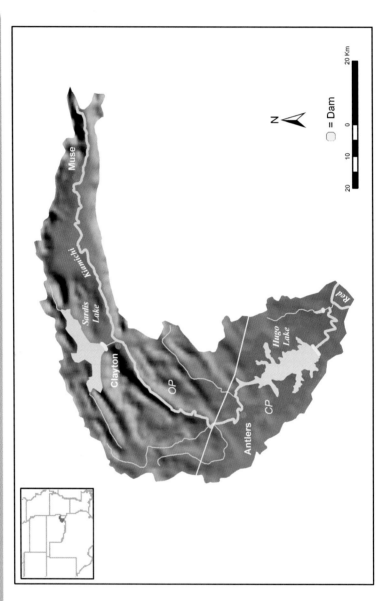

Relief: 701 m
Basin area: 4650 km²
Mean discharge: 48 m³/s
Mean annual precipitation: 110 cm

Mean air temperature: 17°C
Mean water temperature: 16.7°C
No. of fish species: 86
No. of endangered species: 2

Physiographic provinces: Ouachita Province (OP), Coastal Plain (CP)

Major fishes: orangebelly darter, Johnny darter, dusky darter, central stoneroller, bigeye shiner, redfin shiner, rocky shiner, steelcolor shiner, spotted sucker, flathead catfish, smallmouth bass, spotted bass, largemouth bass, blackstriped topminnow, red shiner, gizzard shad, gars, blue sucker, river carpsucker

Major other aquatic vertebrates: snapping turtle, false map turtle, stinkpot turtle, spiny softshell turtle, plain-bellied water snake, cottonmouth, beaver

Major benthic insects: mayflies (*Stenonema, Caenis*), caddisflies (*Oecetis, Nectopsyche*)

Nonnative species: Asian clam, common carp, striped bass in lower main stem, threadfin shad

Major riparian plants: silver maple, box elder, red maple, smooth elder, river birch, blue beech, ash, sweetgum, swamp tupelo, black gum, sycamore, cottonwood, willow oak, American elm, slippery elm

Special features: identified by The Nature Conservancy as one of the most critical watersheds in United States for protecting freshwater biodiversity; population of endangered Ouachita rock pocketbook mussel; reintroduction of river otters

Fragmentation: one major impoundment on main stem; one major impoundment on tributary

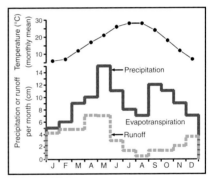

Kiamichi River, Oklahoma (photo by C. C. Vaughn).

Chapter 8

Upper Mississippi River Basin

Michael D. Delong

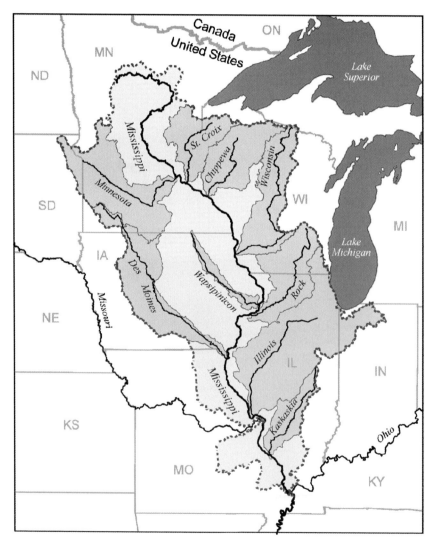

DOI: 10.1016/B978-0-12-375088-4.00008-7

The Upper Mississippi River basin, which represents 10% of the third largest drainage basin in the world, begins as a 1st order stream draining Lake Itasca in the bog and spruce swamps of northern Minnesota and flows south to join the Ohio River as a 10th order alluvial river to form the largest river in North America (see map). The progression of the river from lake outlet to Great River creates an impressive range of physical, chemical, and biological diversity throughout the basin. The Upper Mississippi River basin includes areas within Minnesota, South Dakota, Wisconsin, Illinois, Indiana, Iowa, and Missouri, ranging in latitude from 47°N to 37°N. Because of its generally north-south flow across the temperate zone of North America, climatic conditions vary considerably from its source to the confluence with the Ohio River. Describing the Upper Mississippi River as beginning at Lake Itasca and ending at the confluence of the Ohio River would include the Missouri River basin; however, details on the Missouri River basin are given elsewhere and all information provided for the Upper Mississippi River excludes the Missouri (see Chapter 10).

Archaeological finds suggest that human history in the Upper Mississippi River basin dates back 9000 or more years. Ceremonial and community mounds and other signs of man-made structures found throughout the basin hint at the cultural diversity present in the basin prior to European settlement. The first Europeans credited with exploring the Upper Mississippi, Louis Joliett and Father Jacques Marquette, arrived in 1673. Settlement of the basin began slowly with a few isolated groups in Missouri and Illinois in the early 18th century and in the early 19th century in the northern reaches of the basin. Expansion of settlements came with the advent of the paddle wheeler, which opened the fertile soils of the basin to immigrants seeking to farm their own piece of land. The river today maintains its significant role as a center of commerce for the transportation of goods by barge.

The rivers described in this chapter were selected as representative within each region and to reflect both the common threads among rivers in the Upper Mississippi River basin and their unique attributes. Besides the main stem and the Missouri River, the largest tributary of the Upper Mississippi River is the Illinois River with a basin area of about $75,000\,km^2$, and a mean discharge of $649\,m^3/s$. Other relatively large tributaries include the Wisconsin, Chippewa, Rock, and Des Moines, with mean discharges approaching or exceeding $200\,m^3/s$. Features of the St. Croix, Wapsipinicon, Minnesota, and Kaskaskia rivers are also described.

Upper Mississippi River

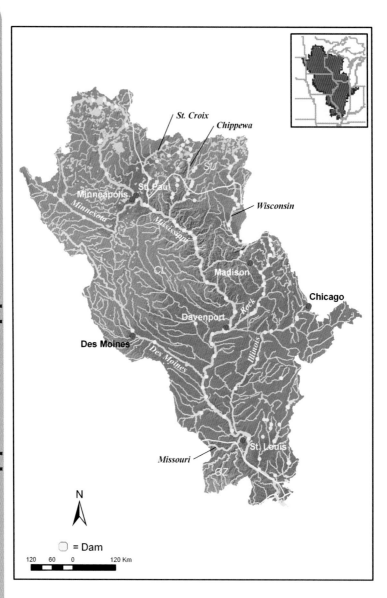

St. Croix
Chippewa
St. Paul
Minneapolis
Minnesota
Wisconsin
Mississippi
CL
Madison
Chicago
Davenport
Rock
Des Moines
Des Moines
Illinois
Missouri
St. Louis

N

○ = Dam

120   60   0   120 Km

Relief: 337 m
Basin area: 489,510 km²
Mean discharge (excluding
Missouri River): 3576 m³/s
Mean annual precipitation: 96 cm

Mean air temperature: 10.5°C
Mean water temperature: 14.3°C
No. of fish species: 145
No. of endangered species: 10

Physiographic provinces: Central Lowland (CL), Superior Upland (SU), Ozark Plateau (OZ)

Major fishes: smallmouth buffalo, shorthead redhorse, river redhorse, gizzard shad, emerald shiner, bluntnose minnow, smallmouth bass, largemouth bass, bluegill, walleye, sauger, channel catfish, flathead catfish, carpsucker, quillback, drum, logperch, paddlefish

Major other aquatic vertebrates: painted turtle, common snapping turtle, smooth softshell turtle, leopard frog, common mudpuppy, northern water snake, muskrat, beaver, river otter

Major benthic insects: Caddisflies (*Hydropsyche, Cheumatopsyche, Potamyia, Cyrnellus, Oecetis, Hydroptila*, mayflies (*Hexagenia, Stenonema, Tricorythodes, Baetisca, Baetis, Caenis*), true flies (*Dicrotendipes, Rheotanytarsus*, stoneflies (*Perlesta*)

Nonnative species: rainbow trout, rainbow smelt, common carp, grass carp, bighead carp, white catfish, ninespine stickleback, striped bass, zebra mussel, Asiatic clam

Major riparian plants: cottonwood, silver maple, river birch, black willow, sandbar willow, box elder, green ash

Special features: large floodplain river that still retains >80% of river–floodplain connectivity

Fragmentation: 11 dams in headwaters; 26 locks and low-head dams on main stem

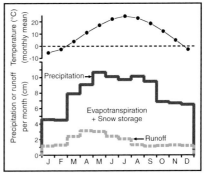

Upper Mississippi River above Little Falls, Minnesota (photo by Tim Palmer).

Minnesota River

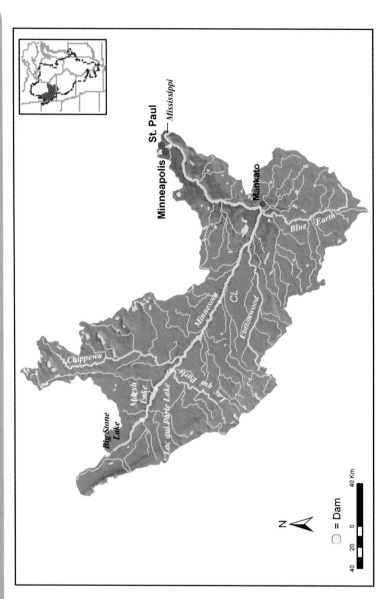

Relief: 85 m

Basin area: 27,030 km²

Mean discharge: 125 m³/s

Mean annual precipitation: 66 cm

Mean air temperature: 7.5°C

Mean water temperature: 10.5°C

No. of fish species: 87

No. of endangered species: 2

Physiographic province: Central Lowland (CL)

Major fishes: shorthead redhorse, quillback, carp, emerald shiner, spotfin shiner, sand shiner, channel catfish, freshwater drum, gizzard shad, bluntnose minnow, smallmouth buffalo, bigmouth buffalo, walleye, fathead minnow, quillback, fathead minnow

Major other aquatic vertebrates: northern leopard frog, mink frog, wood frog, treefrogs, mudpuppy, tiger salamander, painted turtle, common snapping turtle, false map turtle, common map turtle, northern water snake

Major benthic insects: caddisflies (*Hydropsyche*, *Cheumatopsyche*, *Cyrnellus*), mayflies (*Stenonema*, *Stenacron*, *Potamanthus*, *Tricorythodes*), stoneflies (*Isoperla*), true flies (*Glyptotendipes*, *Polypedilum*, *Tanytarsus*)

Nonnative species: brown trout, brook trout, goldfish, common carp, longear sunfish, Asiatic clam

Major riparian plants: cottonwood, green ash, black willow, sandbar willow

Special features: small river in a 1 to 10 km wide channel formed by glacial River Warren

Fragmentation: 6 dams on main stem, including outlets of 3 natural channel lakes

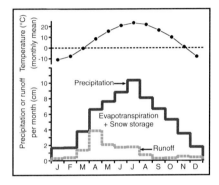

Minnesota River at flood stage, Minnesota (photo by Tim Palmer).

St. Croix River

Eau
Claire
Lakes

St. Croix
Flowage

Snake

Kettle

Gordon

Namekagon

Hinckley

SU

St. Croix

CL

St. Croix
Falls

Stillwater

Lake St. Croix

N

Mississippi

◯ = Dam

30   15   0   30 Km

Relief: 319 m   Mean air temperature: 6.3°C
Basin area: 20,018 km²   Mean water temperature: 10.4°C
Mean discharge: 131 m³/s   No. of fish species: 110
Mean annual precipitation: 78 cm No. of endangered species: 2

Physiographic provinces: Superior Upland (SU), Central Lowland (CL)

Major fishes: emerald shiner, golden redhorse, shorthead redhorse, common carp, gizzard shad, smallmouth bass, bluegill, yellow perch, Johnny darter, spottail shiner, common shiner, spotfin shiner, bluntnose minnow, mimic shiner, brassy minnow, central stoneroller

Major other aquatic vertebrates: painted turtle, common snapping turtle, wood turtle, common map turtle, false map turtle, spiny softshell turtle, northern leopard frog, green frog, spring peeper, treefrogs, tiger salamander, muskrat, beaver, mink

Major benthic insects: caddisflies (*Hydropsyche*, *Ceratopsyche*, *Nectopsyche*, *Lepidostoma*), mayflies (*Baetis*, *Stenonema*, *Stenacron*, *Heptagenia*, *Anthopotamus*, *Siphlonurus*, *Ephemerella*), stoneflies (*Pteronarcys*)

Nonnative species: rainbow trout, brown trout, brook trout, lake trout, rainbow smelt, common carp, zebra mussel, Asiatic clam

Major riparian plants: paper birch, slippery elm, black ash, tamarack, black spruce, white cedar, basswood, red maple, yellow birch

Special features: one of first 8 rivers protected as National Wild and Scenic Rivers; spectacular gorges and

rapids; most pristine river in Upper Mississippi basin

Fragmentation: 2 dams on main stem, but mostly free flowing; 134 small dams on tributaries

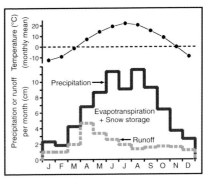

St. Croix River above Grantsburg, Wisconsin (photo by Tim Palmer).

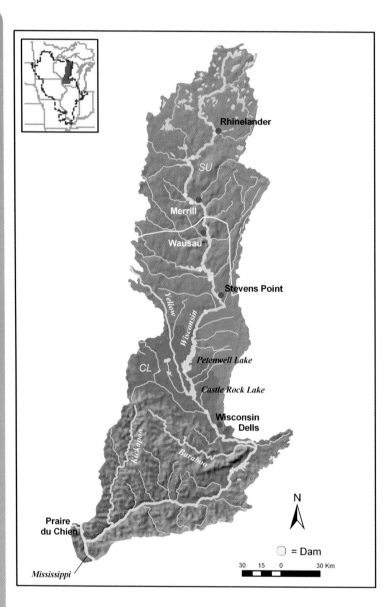

Wisconsin River

Relief: 300 m
Basin area: 30,000 km²
Mean discharge: 261 m³/s
Mean annual precipitation: 85 cm

Mean air temperature: 7.5°C
Mean water temperature: 11.8°C
No. of fish species: 119
No. of endangered species: 2

Physiographic provinces: Central Lowland (CL), Superior Upland (SU)

Major fishes: shorthead redhorse, Johnny darter, logperch, walleye, largemouth bass, smallmouth bass, bluegill, yellow perch, river carpsucker, smallmouth buffalo, gizzard shad, freshwater drum

Major other aquatic vertebrates: northern water snake, common snapping turtle, common musk turtle, wood turtle, painted turtle, common map turtle, Ouachita map turtle, smooth softshell turtle, green frog, mink frog, muskrat, beaver

Major benthic insects: caddisflies (*Cheumatopsyche*, *Hydropsyche*, *Potamyia*, *Nectopsyche*), mayflies (*Stenonema*, *Baetisca*, *Baetis*, *Caenis*, *Stenacron*, *Procloeon*, *Isonychia*), stoneflies (*Isoperla*)

Nonnative species: rainbow trout, brown trout, common carp, grass pickerel

Major riparian plants: silver maple, river birch, swamp white oak, green ash, cottonwood, black willow

Special features: largest river in Wisconsin; Grandfather Falls rapids descend 28 m over 2.4 km

Fragmentation: heavily impounded in middle and upper reaches

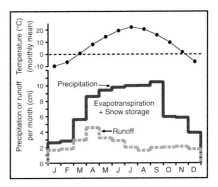

Wisconsin River at the Wisconsin Dells (photo by A.C. Benke).

Illinois River

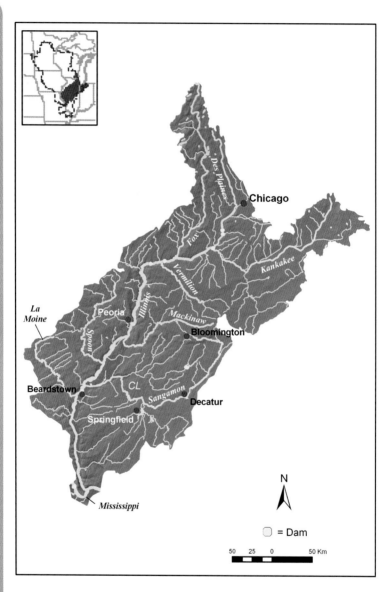

Relief: 45 m
Basin area: 75,136 km²
Mean discharge: 649 m³/s
Mean annual precipitation: 92 cm

Mean air temperature: 10.4°C
Mean water temperature: 16°C
No. of fish species: 127
No. of endangered species: 2

Physiographic province: Central Lowland (CL)

Major fishes: gizzard shad, emerald shiner, freshwater drum, white bass, bluegill, green sunfish, largemouth bass, common carp, smallmouth buffalo, white sucker, bluntnose minnow, channel catfish, flathead catfish, bowfin, shortnose gar, grass pickerel, quillback, carpsucker

Major other aquatic vertebrates: common snapping turtle, smooth softshell turtle, common map turtle, false map turtle, slider, treefrogs, southern leopard frog, tiger salamander, northern water snake

Major benthic insects: caddisflies (*Hydropsyche*, *Cheumatopsyche*, *Potamyia*, *Cyrnellus*), mayflies (*Stenonema*, *Baetis*, *Hexagenia*, *Heptagenia*), true flies (*Robackia*, *Rheosmittia*)

Nonnative species: rainbow trout, rainbow smelt, common carp, grass carp, bighead carp, silver carp, zebra mussel, Asiatic clam, round goby

Major riparian plants: silver maple, ash, box elder, black willow, hackberry

Special features: lower river flows through channel

abandoned by Upper Mississippi; broad low-gradient floodplain results in protracted spring flood

Fragmentation: 5 low-head navigation dams on main stem

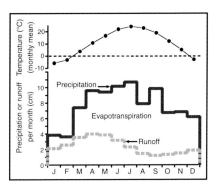

Upper Illinois River at Starved Rock State Park near Utica, Illinois.

Chippewa River

Relief: 290 m
Basin area: 24,827 km^2
Mean discharge: 218 m^3/s
Mean annual precipitation: 80 cm

Mean air temperature: 6°C
Mean water temperature: 11°C
No. of fish species: 110
No. of endangered species: 5

Physiographic provinces: Central Lowland (CL), Superior Upland (SU)

Major fishes: shorthead redhorse, golden redhorse, smallmouth buffalo, carpsucker, common carp, mooneye, gizzard shad, shovelnose sturgeon, smallmouth bass, northern pike, channel catfish, walleye, sauger, muskellunge, emerald shiner, paddlefish

Major other aquatic vertebrates: muskrat, beaver, river otter, snapping turtle, common map turtle, false map turtle, spiny softshell turtle, leopard frog, green frog, treefrogs, northern water snake

Major benthic insects: mayflies (*Pseudocloeon*, *Stenonema*, *Hexagenia*), stoneflies (*Allocapnia*), caddisflies (*Cheumatopsyche*, *Hydropsyche*, *Agraylea*, *Cyrnellus*), true flies (*Dasyhelia*, *Cladotanytarsus*, *Dicrotendipes*, *Orthocladius*, *Robackia*, *Simulium*)

Nonnative species: rainbow trout, brown trout, rainbow smelt, common carp, curly-leaved pondweed, Eurasian watermilfoil

Major riparian plants: cottonwood, silver maple, black willow, green ash, American elm, river birch, white swamp oak

Special features: sand carried by Chippewa

River led to formation of Lake Pepin on Upper Mississippi River

Fragmentation: over 125 dams, including 16 impoundments $>12\,km^2$; hydrology extensively altered by hydropower

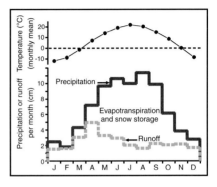

Chippewa River, Wisconsin (photo by Tim Palmer).

Wapsipinicon River

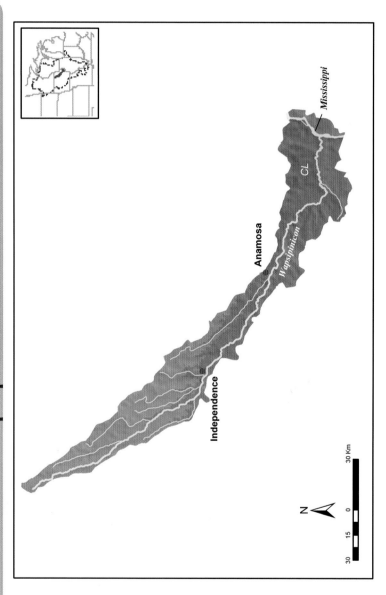

Relief: 195 m
Basin area: 6050 km²
Mean discharge: 47 m³/s
Mean annual precipitation: 88 cm

Mean air temperature: 9°C
Mean water temperature: 12°C
No. of fish species: 74
No. of endangered species: 1

Physiographic province: Central Lowland (CL)

Major fishes: American brook lamprey, spotfin shiner, Mississippi silvery minnow, bigmouth shiner, sand shiner, bluntnose minnow, bullhead minnow, river carpsucker, shorthead redhorse, Johnny darter

Major other aquatic vertebrates: muskrat, beaver, river otter, snapping turtle, western painted turtle, common map turtle, false map turtle, mudpuppy, spiny softshell turtle, treefrogs, leopard frog

Major benthic insects: caddisflies (*Hydropsyche, Cheumatopsyche, Ceratopsyche, Nectopsyche*), mayflies (*Tricorythodes, Heptagenia, Baetis, Caenis, Stenonema*), stoneflies (*Paragnetina, Pteronarcys*), true flies (*Ablabesmyia, Cricotopus/ Orthocladius, Simulium, Atherix*)

Nonnative species: common carp, rainbow trout, brown trout

Major riparian plants: silver maple, cottonwood, black willow, sandbar willow

Special features: one of few Iowa rivers that retains mature stands of riparian and floodplain vegetation; very narrow basin

Fragmentation: 21 dams, mostly small impoundments on tributaries of upper basin

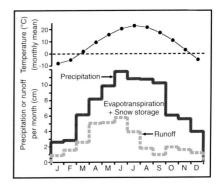

Wapsipinicon River north of Tripoli, Iowa (photo by S. Porter, U.S. Geological Survey).

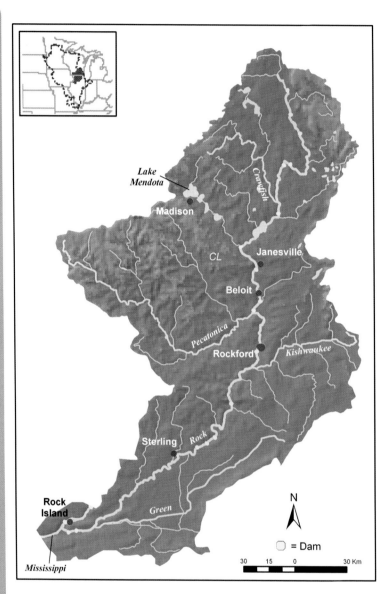

Rock River

Relief: 155 m
Basin area: 28,101 km²
Mean discharge: 184 m³/s
Mean annual precipitation: 99 cm

Mean air temperature: 10°C
Mean water temperature: 12°C
No. of fish species: 115
No. of endangered species: 1

Physiographic province: Central Lowland (CL)

Major fishes: gizzard shad, carp, spotfin shiner, smallmouth buffalo, largemouth buffalo, channel catfish, white bass, smallmouth bass, walleye, northern pike, freshwater drum, white sucker, green sunfish, Johnny darter, central stoneroller

Major other aquatic vertebrates: tiger salamander, common mudpuppy, bullfrog, green frog, northern leopard frog, pickerel frog, treefrogs, common snapping turtle, common map turtle, painted turtle, slider, common musk turtle, northern water snake

Major benthic insects: caddisflies (*Hydropsyche*, *Cheumatopsyche*, *Nectopsyche*), mayflies (*Stenonema*, *Stenacron*, *Baetis*), true flies (*Simulium*)

Nonnative species: rainbow trout, brown trout, lake trout, common carp, grass carp, goldfish, rusty crayfish, zebra mussel, Eurasian watermilfoil

Major riparian plants: cottonwood, silver maple, box elder, black willow, sandbar willow

Special features: upper river drains unique glacial formations, creating constrained channels for many tributaries

Fragmentation: 19 hydroelectric dams on main stem and 272 dams throughout the basin

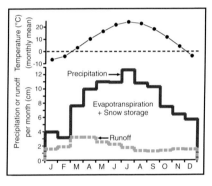

Rock River north of Watertown, Wisconsin (photo by S. Wade).

Des Moines River

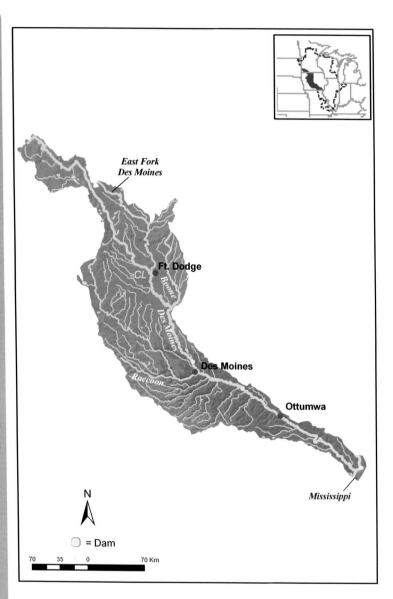

East Fork
Des Moines

Ft. Dodge

CL

Boone

Des Moines

Raccoon

Des Moines

Ottumwa

Mississippi

N

○ = Dam

70   35   0   70 Km

Relief: 290 m   Mean air temperature: 11°C
Basin area: 31,127 km²   Mean water temperature: 11.6°C
Mean discharge: 182 m³/s   No. of fish species: 84
Mean annual precipitation: 96 cm No. of endangered species: 3

Physiographic province: Central Lowland (CL)

Major fishes: shorthead redhorse, common carp, smallmouth buffalo, emerald shiner, bluntnose minnow, bluegill, largemouth bass

Major other aquatic vertebrates: snapping turtle, painted turtle, spiny softshell turtle, smooth softshell turtle, red-eared slider, treefrogs, mudpuppy, leopard frog, pickerel frog, green frog, muskrat

Major benthic insects: caddisflies (*Hydropsyche, Cheumatopsyche, Ceratopsyche, Nectopsyche*), mayflies (*Tricorythodes, Heptagenia, Baetis, Caenis, Stenonema*), stoneflies (*Paragnetina, Pteronarcys*), true flies (*Ablabesmyia, Cricotopus/ Orthocladius, Simulium*)

Nonnative species: common carp, grass carp, bighead carp, striped bass

Major riparian plants: cottonwood, ash, black willow, sandbar willow

Special features: largest river in Iowa, with drainage basin covering 23% of state

Fragmentation: 2 major impoundments ($>12\,\mathrm{km}^3$) and 58 small to medium-size impoundments on main stem and tributaries

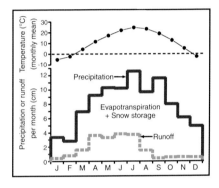

Des Moines River near Des Moines, Iowa (photo by Tim Palmer).

Kaskaskia River

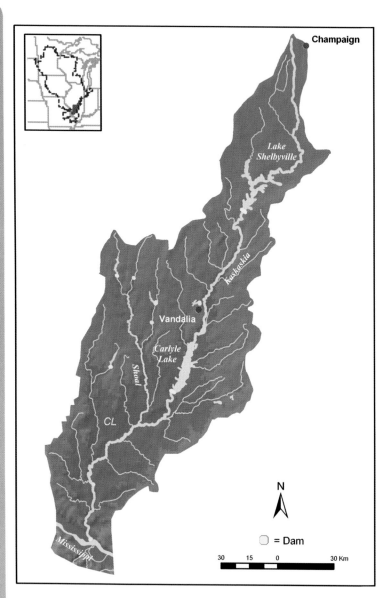

Relief: 100 m
Basin area: 15,025 km²
Mean discharge: 107 m³/s
Mean annual precipitation: 99 cm

Mean air temperature: 11°C
Mean water temperature: 15.2°C
No. of fish species: 112
No. of endangered species: 1

Physiographic province: Central Lowland (CL)

Major fishes: common carp, shorthead redhorse, channel catfish, freshwater drum, bluegill, largemouth bass, flathead catfish, white crappie, yellow bass, white bass, gizzard shad, sand shiner, bigmouth buffalo, western mosquitofish

Major other aquatic vertebrates: smooth softshell turtle, painted turtle, false map turtle, river cooter, smooth softshell turtle, slider, southern leopard frog, pickerel frog, green frog, treefrogs, tiger salamander, smallmouth salamander, northern water snake, diamondback water snake

Major benthic insects: caddisflies (*Hydropsyche*, *Cheumatopsyche*), mayflies (*Stenonema*, *Caenis*, *Tricorythodes*), stoneflies (*Taeniopteryx*), true flies (Atherix)

Nonnative species: bighead carp, common carp, silver carp, goldfish, white catfish, redear sunfish, zebra mussel

Major riparian plants: silver maple, box elder, black willow, cottonwood

Special features: southernmost large tributary of upper Mississippi; largest contiguous tracts of forest, including bottomland, in Illinois

Fragmentation: 107 dams on the main stem tributaries, including 4 impoundments $>12\,km^2$; lock and dam near mouth creates 21 km long pool

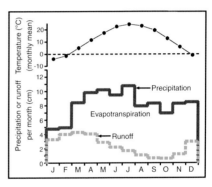

Kaskaskia River, showing Shelbyville Lake and Dam near Shelbyville, Illinois (photo by USACE).

Chapter 9

Ohio River Basin

David White, Karla Johnston, and Michael Miller

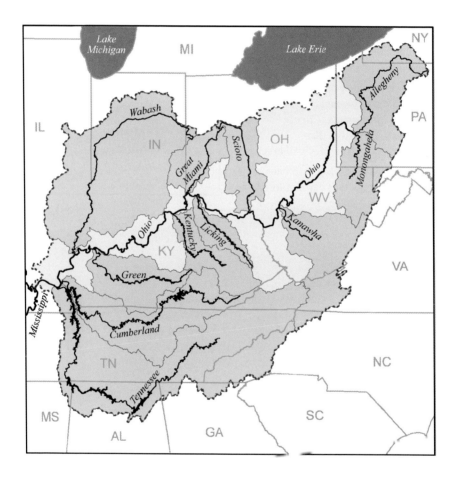

The word "Ohio" comes from Iroquois meaning "beautiful river." Early French explorers also called it "La Belle Rivière." The Ohio basin is the third largest by discharge (8733 m^3/s) in the United States, accounting for more than 40% of the discharge of the Mississippi River, while making up only 16% of its drainage area. The basin lies between 34° and 41°N latitude and 77° and 89°W longitude and drains major portions of eight states and minor parts of six additional states (528,200 km^2) from New York in the northeast to Alabama in the south to Illinois in the west (see map). The eastern portion of the basin has its tributary origins in the Appalachian Mountains, and the northern tributaries border on Laurentian Great Lakes drainages. Climate is continental with abundant rainfall, cool moist winters, and warm summers. With the exception of some prairie in the north and west, the region historically was heavily forested, but today agriculture is a major feature along with many large urban areas. Patterns of glaciation, mountains, and varied geology have resulted in a highly diverse biological environment.

Humans have lived along the river for at least the past 12,000 years. Paleo-Indian foraging cultures were present from 11,500 to 10,000 years ago, Archaic cultures followed from 10,000 to 3000 years ago and then Woodland cultures between 3000 and 10,000 years ago. At the time of the first French explorers, the basin was home to a wide variety of Native Americans including Shawnee, Mosopelea, Erie-Iroquois, Cherokee, and Miami-Pottawatomie cultures. French and British traders competed with each other for the lucrative fur trade for more than a century. In the 1780s, European settlers poured in from the south through Georgia, east through the Cumberland Gap, and down the Ohio River. Within 30 years, they had completely displaced Native Americans and had begun to permanently alter the landscape. Development was rapid; all the states in the Ohio basin had been admitted to the Union by 1818, and most all of the major cities had been established.

In this chapter, we describe the Ohio River main stem and several tributaries that reflect the basin's diversity. The Tennessee River is the southernmost and largest tributary, with a basin area of almost 106,000 km^2 and a mean discharge of about 2000 m^3/s. Several other very large tributaries with mean discharges >500 m^3/s include the Wabash, Cumberland, Kanawha, and Allegheny. Also described are the Green, Kentucky, Great Miami, Licking, Scioto, and Monongahela.

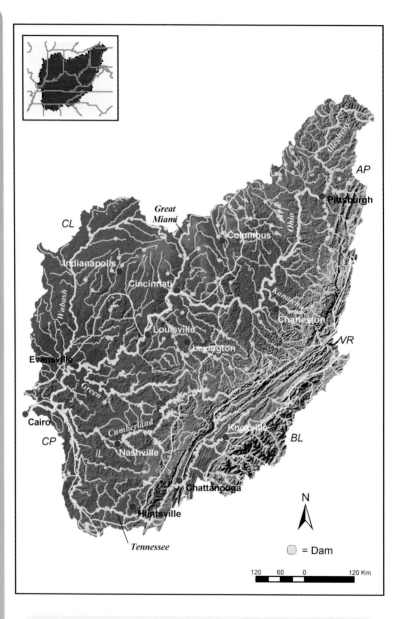

Ohio River

Relief: 2300 m
Basin area: 529,000 km²
Mean discharge: 8733 m³/s
Mean annual precipitation: 104 cm

Mean air temperature: 12°C
Mean water temperature: 14°C
No. of fish species: 240 to 250
No. of endangered species: 8

Physiographic provinces: Blue Ridge (BL), Valley and Ridge (VR), Appalachian Plateaus (AP), Central Lowland (CL), Interior Low Plateaus (IL), Coastal Plain (CP)

Major fishes: gizzard shad, skipjack herring, emerald shiner, white sucker, golden redhorse, black buffalo, river carpsucker, channel catfish, flathead catfish, longnose gar, shortnose gar, largemouth bass, smallmouth bass, bluegill, white crappie, freshwater drum

Major other aquatic vertebrates: snapping turtle, stinkpot, common map turtle, common water snake, queen snake, green frog, bullfrog, bald eagle, osprey, great blue heron, Canada goose, beaver, muskrat, river otter

Major benthic insects: mayflies (*Hexagenia, Ephemerella, Caenis, Stenacron*), stoneflies (*Isoperla*), caddisflies (*Hydroptila, Hydropsyche, Ceraclea, Cyrnellus, Polycentropus, Potamyia, Chimarra*), true flies (*Tipula*)

Nonnative species: zebra mussel, Asiatic clam, yellow perch, common carp, goldfish, rainbow smelt, fathead chub, rosefin shiner, bighead carp, striped bass, brown trout, brittle naiad, curly pondweed, Eurasian watermilfoil

Major riparian plants: red maple, cottonwood, black willow, sycamore, black gum, sugar hackberry, water willow, buttonbush

Special features: high biodiversity, especially fishes, freshwater mussels, crayfishes, and aquatic insects

Fragmentation: 20 low-water locks and dams on main stem; >700 major dams in basin

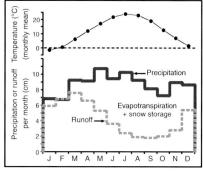

Ohio River at Paducah, Kentucky (photo by G. Harris).

Tennessee River

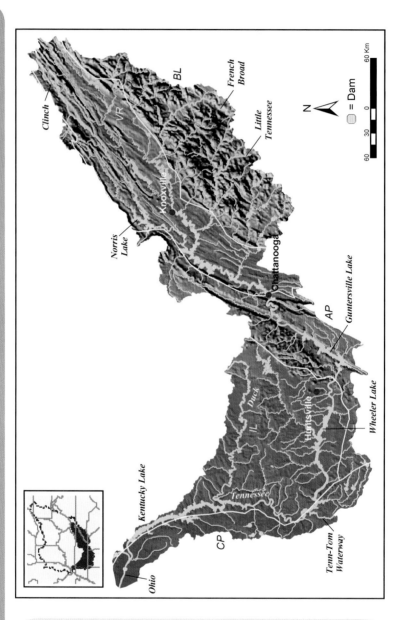

Relief: 1910 m
Basin area: 105,870 km²
Mean discharge: 2000 m³/s
Mean annual precipitation: 105 cm

Mean air temperature: 13°C
Mean water temperature: 19°C
No. of fish species: 225 to 240
No. of endangered species: 31

Physiographic provinces: Valley and Ridge (VR), Blue Ridge (BL), Appalachian Plateaus (AP), Interior Low Plateaus (IL), Coastal Plain (CP)

Major fishes: gizzard shad, threadfin shad, largemouth bass, channel catfish, white bass, striped bass, smallmouth bass, freshwater drum, paddlefish, white crappie, bluegill, flathead catfish, white sucker, spotfin shiner, striped shiner, emerald shiner, orangethroat darter, fantail darter, sauger

Major other aquatic vertebrates: snapping turtle, stinkpot, mud turtle, common map turtle, midland painted turtle, spiny softshell turtle, sliders, common water snake, queen snake, green frog, bullfrog, cottonmouth, bald eagle, osprey, great blue heron, coot, mallard, common loon, ring-billed gull, American white pelican, beaver, river otter, muskrat

Major benthic insects: mayflies (*Hexagenia*, *Caenis*), stoneflies (*Acroneuria*), caddisflies (*Hydroptila*, *Cheumatopsyche*, *Ceraclea*, *Brachycentrus*)

Nonnative species: *Daphnia lumholtzi*, Asiatic clam, zebra mussel, goldfish, grass carp, tench, golden shiner, rainbow smelt, Ohrid trout, rainbow trout, brown trout, yellow perch, bighead carp, brook stickleback, Eurasian watermilfoil, purple loosestrife, parrot-feather,

alligatorweed, brittle naiad, curly pondweed, watercress, Nepal grass

Major riparian plants: water willow, red maple, buttonbush, cottonwood, black gum, American sycamore, black willow

Special features: most endemic fishes, mussels, and crayfishes of any river in North America; Holston, Clinch, North Fork, Duck, and Powel tributaries have retained much diversity; Obed is National Wild and Scenic River

Fragmentation: 48 multipurpose dams on main stem and major tributaries

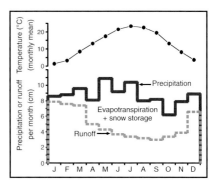

Tennessee River near Reidland, Kentucky (photo by G. Harris).

Cumberland River

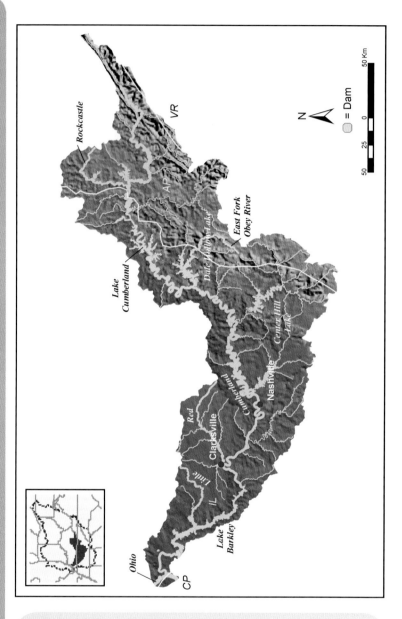

Relief: 1160 m
Basin area: 46,430 km²
Mean discharge: 862 m³/s
Mean annual precipitation: 127 cm

Mean air temperature: 14°C
Mean water temperature: 16°C
No. of fish species: 172 to 186
No. of endangered species: 19

Physiographic provinces: Valley and Ridge (VR), Interior Low Plateaus (IL), Appalachian Plateaus (AP)

Major fishes: gizzard shad, threadfin shad, channel catfish, largemouth bass, smallmouth bass, spotted bass, white bass, striped bass, common carp, white sucker, freshwater drum, shortnose gar, longnose gar, bluegill, green sunfish, white crappie, paddlefish

Major other aquatic vertebrates: common snapping turtle, stinkpot, mud turtle, common map turtle, midland painted turtle, spiny softshell turtle, common water snake, queen snake, mudpuppy, plethodontid salamanders, bullfrog, green frog, bald eagle, osprey, great blue heron, coot, mallard, ring-billed gull, beaver, muskrat, river otter

Major benthic insects: mayflies (*Stenonema, Caenis, Tricorythodes, Hexagenia*), caddisflies (*Cheumatopsyche, Ceraclea, Hydroptila*)

Nonnative species: Asiatic clam, zebra mussel, *Daphnia lumholtzi*, common carp, goldfish, striped bass, rainbow trout, brown trout, yellow perch, bighead carp, grass carp, tench, grass carp, golden shiner, alewife, Eurasian watermilfoil, hydrilla, brittle naiad, watercress, curly pondweed, alligatorweed, yellow iris, Uruguay seedbox, parrot-feather

Major riparian plants: water willow, buttonbush, red maple, cottonwood, American sycamore, black willow, black gum

Special features: Cumberland Falls; Rockcastle River with many endemic fishes and mussels; caves and karst flow in midregion

Fragmentation: 10 major dams and many tributary dams

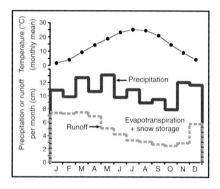

Cumberland River near Smithland, Kentucky (photo by G. Harris).

Wabash River

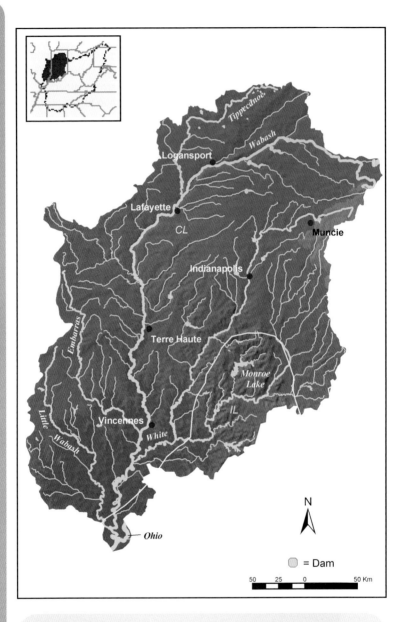

= Dam

50   25   0   50 Km

Relief: 275 m
Basin area: 85,340 km²
Mean discharge: 1001 m³/s
Mean annual precipitation: 96 cm

Mean air temperature: 11°C
Mean water temperature: 15°C
No. of fish species: >95
No. of endangered species: 3

Physiographic provinces: Central Lowland (CL), Interior Low Plateaus (IL)

Major fishes: gizzard shad, common carp, steelcolor shiner, spotfin shiner, channel catfish, longnose gar, shortnose gar, quillback, river carpsucker, shorthead redhorse, silver redhorse, golden redhorse, freshwater drum, emerald shiner, longear sunfish, white crappie, sauger, white bass, shovelnose sturgeon

Major other aquatic vertebrates: snapping turtle, stinkpot, midland painted turtle, spiny softshell turtle, common water snake, red-bellied water snake, Kirtland's water snake, queen snake, newt, mudpuppy, leopard frog, bullfrog, green frog, bald eagle, osprey, belted kingfisher, mallard, coot, beaver, muskrat, river otter

Major benthic insects: mayflies (*Hexagenia, Caenis, Brachycerus, Tricorythodes, Stenonema, Heptagenia*), caddisflies (*Cheumatopsyche, Hydropsyche, Ceraclea*), stoneflies (*Taeniopteryx, Allocapnia, Isoperla, Perlesta*)

Nonnative species: Asiatic clam, zebra mussel, common carp, goldfish, bighead carp, yellow perch, brook trout, rainbow trout,

purple loosestrife, Eurasian watermilfoil

Major riparian plants: red maple, cottonwood, American sycamore, black willow, black gum, sugar hackberry, buttonbush

Special features: 267 km long Tippecanoe River remains relatively pristine, drains several northern glacial pothole lakes, and has extensive aquatic macrophytes beds

Fragmentation: one major dam at Rkm 662; free flowing from there to confluence with Ohio River

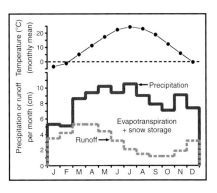

Wabash River near Montezuma, Indiana (photo by T. Harris).

Kanawha River

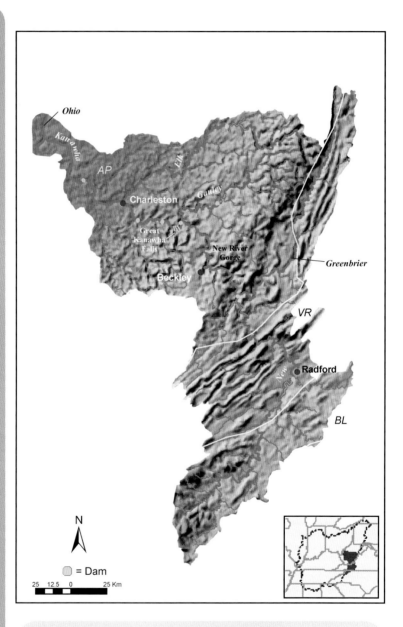

Relief: 1545 m
Basin area: 31,690 km²
Mean discharge: 537 m³/s
Mean annual precipitation: 94 cm

Mean air temperature: 13°C
Mean water temperature: 14°C
No. of fish species: 126
No. of endangered species: 3

Physiographic provinces: Blue Ridge (BL), Valley and Ridge (VR), Appalachian Plateaus (AP)

Major fishes: emerald shiner, spotfin shiner, white sucker, river redhorse, golden redhorse, channel catfish, flathead catfish, longnose gar, white bass, smallmouth bass, green sunfish, longear sunfish, spotted bass, white crappie, sauger, sharpnose darter, Kanawha darter, Appalachia darter, Kanawha minnow, New River shiner, bigmouth chub

Major other aquatic vertebrates: snapping turtle, stinkpot, midland painted turtle, hieroglyphic river cooter, spiny softshell turtle, common water snake, queen snake, bullfrog, green frog, hellbender, newt, mudpuppy, osprey, bald eagle, mallard, coot, great blue heron, beaver, muskrat

Major benthic insects: mayflies (*Stenonema, Isonychia, Leptophlebia*), stoneflies (*Allocapnia, Taeniopteryx, Isoperla, Peltoperla*), caddisflies (*Hydropsyche, Ceraclea, Hydroptila, Brachycentrus*), true flies (*Tipula*)

Nonnative species: Asiatic clam, zebra mussel, rusty crayfish, alewife, common carp, goldfish, golden shiner, striped bass, fathead minnow, tench, purple loosestrife, brittle naiad, curly pondweed, yellow iris

Major riparian plants: red maple, cottonwood, black willow, American sycamore, speckled alder, tulip poplar

Special features: Great Kanawha Falls limits species migration; New is oldest North American river; Greenbrier is unimpounded; New and Greenbrier have generally good water quality and significant free-flowing stretches

Fragmentation: 4 locks and dams on Kanawha main stem, 1 dam on Elk and Gurley rivers, 2 dams on New River

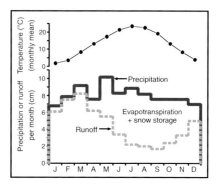

Kanawha River northeast of Charleston, West Virginia (photo by J. Boynton).

Green River

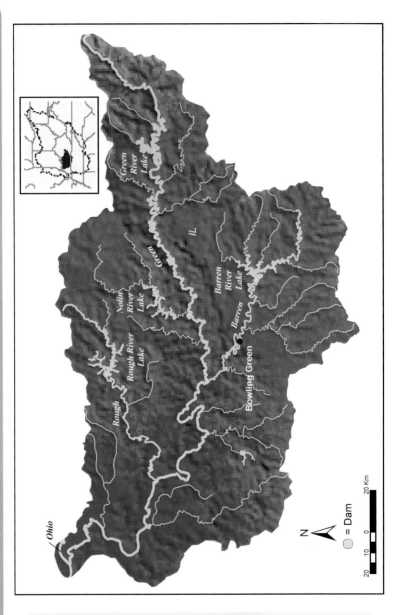

Relief: 385 m
Basin area: 23,850 km²
Mean discharge: 420 m³/s
Mean annual precipitation: 118 cm

Mean air temperature: 14°C
Mean water temperature: 16°C
No. of fish species: 151
No. of endangered species: >31

Physiographic province: Interior Low Plateaus (IL)

Major fishes: white sucker, gizzard shad, bluegill, rock bass, largemouth bass, spotted bass, splendid darter, orangefin darter, teardrop darter, Kentucky snubnose darter, blackfin sucker, channel catfish, cavefish, spotted sunfish, flier

Major other aquatic vertebrates: banded water snake, slider, false map turtle, stinkpot, spiny softshell turtle, mudpuppy, hellbender, bald eagle, osprey, great blue heron, beaver, muskrat, river otter

Major benthic insects: mayflies (*Hexagenia, Caenis, Ephemerella, Isonychia, Stenonema*), stoneflies (*Allocapnia, Acroneuria, Isoperla, Taeniopteryx, Neoperla*), caddisflies (*Ceraclea, Cheumatopsyche, Hydropsyche, Hydroptila*)

Nonnative species: Asiatic clam, zebra mussel, common carp, goldfish, striped bass, yellow perch, fathead minnow, brown trout, rainbow trout, yellow iris, parrot-feather, brittle naiad, watercress, Eurasian watermilfoil, purple loosestrife

Major riparian plants: red maple, buttonbush, black willow, cottonwood, sycamore, water willow, cattails

Special features: upper portion of river lies in karst topography with numerous sinkholes and caves, including Mammoth Cave

Fragmentation: 13 major dams, 7 of which have navigation locks for barge traffic

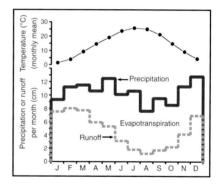

Green River near Rockport, Kentucky (photo by J. Boynton).

Kentucky River

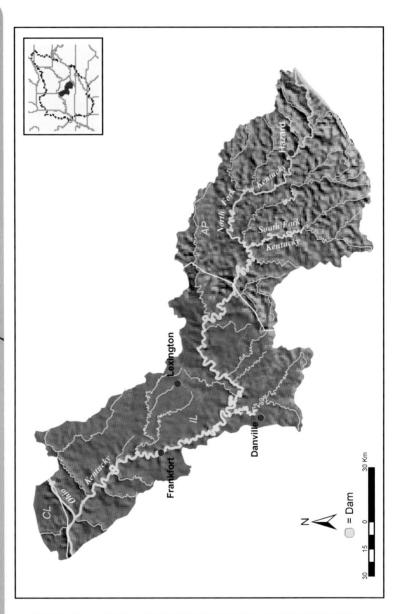

Relief: 840 m
Basin area: 18,025 km²
Mean discharge: 285 m³/s
Mean annual precipitation: 111 cm

Mean air temperature: 13°C
Mean water temperature: 15°C
No. of fish species: 110 to 115
No. of endangered species: 13

Physiographic provinces: Appalachian Plateaus (AP), Interior Low Plateaus (IL)
Major fishes: gizzard shad, bluegill, rock bass, largemouth bass, spotted bass, redside dace, mimic shiner, eastern sand darter, slender chub, sharpnose darter, channel catfish, spotted sucker, golden redhorse, shorthead redhorse, silver redhorse, striped shiner, longnose gar, rainbow darter, greenside darter, sauger
Major other aquatic vertebrates: banded water snake, midland painted turtle, stinkpot, spiny softshell turtle, mudpuppy, hellbender, bullfrog, green frog, bald eagle, great blue heron, beaver, muskrat
Major benthic insects: mayflies (*Ephemerella*, *Isonychia*), stoneflies (*Allocapnia*, *Acroneuria*, *Isoperla*, *Taeniopteryx*, *Perlesta*), caddisflies (*Ceraclea*, *Cheumatopsyche*, *Hydropsyche*)
Nonnative species: Asiatic clam, zebra mussel, common carp, goldfish, striped bass, yellow perch, brook stickleback, rainbow trout, brown trout, purple loosestrife, brittle naiad, curly pondweed
Major riparian plants: speckled alder, red maple, buttonbush, black willow, water willow, cottonwood, American sycamore
Special features: middle portion of river passes through Daniel Boone National Forest, where invertebrate and fish diversity are high; portions of Red tributary, including Red River Gorge, are in National Wild and Scenic River system
Fragmentation: 14 navigation dams on main stem

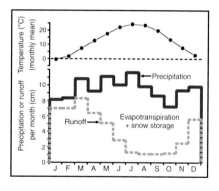

Kentucky River near Tyrone, Kentucky (photo by J. Boynton).

Great Miami River

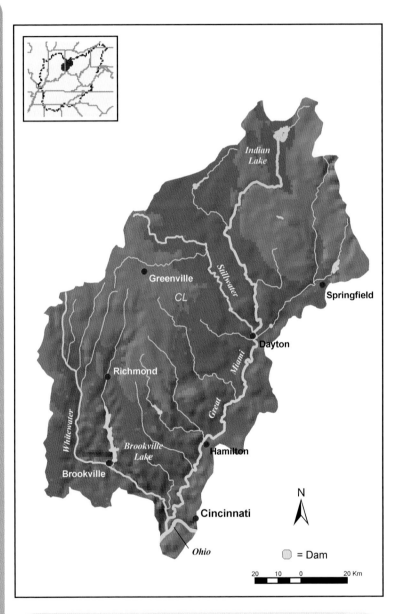

Relief: 305 m
Basin area: 13,915 km²
Mean discharge: 152 m³/s
Mean annual precipitation: 102 cm

Mean air temperature: 11°C
Mean water temperature: 15°C
No. of fish species: 120 to 125
No. of endangered species: 1

Physiographic province: Central Lowland (CL)

Major fishes: gizzard shad, stoneroller, white sucker, hog sucker, rock bass, largemouth bass, spotted bass, smallmouth bass, mimic shiner, creek chub, sharpnose darter, channel catfish, bullhead, golden redhorse, rainbow darter, black crappie, sauger, green sunfish, spotted sucker, striped shiner, spotfin shiner, black buffalo, striped bass

Major other aquatic vertebrates: common water snake, queen snake, midland painted turtle, stinkpot, spiny softshell turtle, mudpuppy, green frog, pickerel frog, bullfrog, wood duck, great blue heron, mallard, beaver, muskrat

Major benthic insects: mayflies (*Isonychia, Stenacron, Ephemerella, Stenonema*), stoneflies (*Allocapnia, Isoperla, Acroneuria*), caddisflies (*Hydroptila, Hydropsyche, Cheumatopsyche, Ceratopsyche, Ceraclea*), true flies (*Tipula*)

Nonnative species: Asiatic clam, zebra mussel, common carp, goldfish, striped bass, rainbow smelt, tench, brown trout, rainbow trout, purple loosestrife, Eurasian watermilfoil, European brooklime

Major riparian plants: red maple, cottonwood, sycamore, black willow, water willow

Special features: Great Miami Buried Valley aquifer extends from Dayton to Ohio River and roughly follows pathway of river; Stillwater River designated as exceptional warmwater habitat

Fragmentation: minor; Taylorsville dam north of Dayton

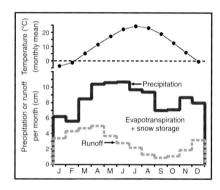

Great Miami River at Miamitown, Ohio (photo by K. Wilhelm).

Licking River

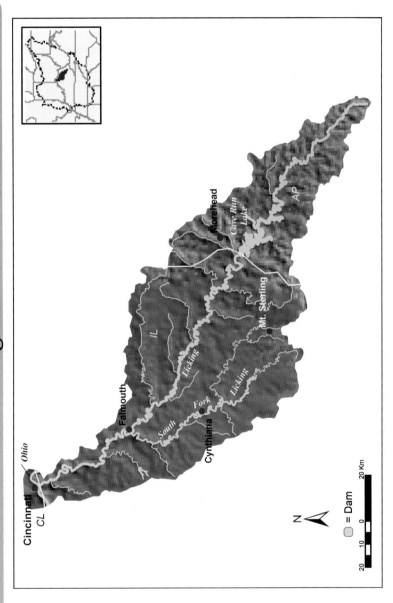

Relief: 345 m
Basin area: 9600 km^2
Mean discharge: 145 m^3/s
Mean annual precipitation: 112 cm

Mean air temperature: 12°C
Mean water temperature: 14°C
No. of fish species: 110
No. of endangered species: 13

Physiographic provinces: Appalachian Plateaus (AP), Interior Low Plateaus (IL)

Major fishes: rock bass, largemouth bass, spotted bass, redside dace, mimic shiner, eastern sand darter, slender chub, sharpnose darter, channel catfish, white sucker, spotted sucker, golden redhorse, striped shiner, longnose gar, fantail darter, rainbow darter, greenside darter, Johnny darter, sauger, white crappie, black crappie, longear sunfish

Major other aquatic vertebrates: banded water snake, midland painted turtle, stinkpot, spiny softshell turtle, mudpuppy, hellbender, newt, bullfrog, green frog, bald eagle, great blue heron, belted kingfisher, beaver, river otter, muskrat

Major benthic insects: mayflies (*Baetis, Caenis, Ephemerella, Isonychia*), stoneflies (*Allocapnia, Acroneuria, Isoperla, Taeniopteryx, Perlesta*), caddisflies (*Ceraclea, Hydropsyche, Hydroptila*), true flies (*Tipula*)

Nonnative species: Asiatic clam, zebra mussel, grass carp, common carp, goldfish, rainbow trout, striped bass, brook stickleback, brittle naiad

Major riparian plants: speckled alder, red maple, buttonbush, black willow, water willow, American sycamore, cottonwood

Special features: middle portion passes through Daniel Boone National Forest, where invertebrate and fish diversity is high

Fragmentation: one major dam at Rkm 278, low-water dams on several tributaries

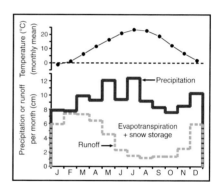

Licking River near Midland, Kentucky (photo by J. Boynton).

Scioto River

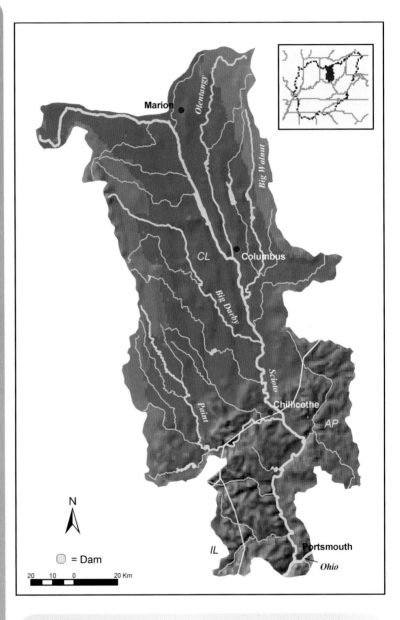

Relief: 307 m
Basin area: 16,882 km²
Mean discharge: 189 m³/s
Mean annual precipitation: 98 cm

Mean air temperature: 11°C
Mean water temperature: 15°C
No. of fish species: 120 to 130
No. of endangered species: 3

Physiographic provinces: Central Lowland (CL), Appalachian Plateaus (AP), Interior Low Plateaus (IL)

Major fishes: sand shiner, silver redhorse, green sunfish, rock bass, smallmouth bass, gizzard shad, mottled sculpin, flathead catfish, rock bass, spotted bass, redside dace, mimic shiner, eastern sand darter, slender chub, sharpnose darter, spotted sucker, hog sucker, golden redhorse, spotfin shiner, striped shiner, longnose gar, rainbow darter, greenside darter, Johnny darter, sauger, black crappie

Major other aquatic vertebrates: snapping turtle, stinkpot, midland painted turtle, spiny softshell turtle, common water snake, queen snake, bullfrog, green frog, pickerel frog, bald eagle, great blue heron, osprey, beaver, river otter, common loon, mallard, coot, muskrat

Major benthic insects: mayflies (*Acerpenna, Stenacron, Isonychia, Hexagenia*), stoneflies (*Allocapnia, Isoperla, Perlesta, Acroneuria*), caddisflies (*Hydroptila, Hydropsyche, Cheumatopsyche*), true flies (*Tipula, Simulium*)

Nonnative species: Asiatic clam, zebra mussel, tench, white bass, common carp, goldfish, yellow iris, purple loosestrife, Eurasian watermilfoil,

brittle naiad, curly pondweed, European brooklime

Major riparian plants: red maple, cottonwood, American sycamore, black willow, black gum, buttonbush, water willow

Special features: river from Chillicothe downstream in Appalachian Plateaus is heavily forested; Big and Little Darby creeks are National Wild and Scenic Rivers

Fragmentation: one major dam on main stem, several tributary dams

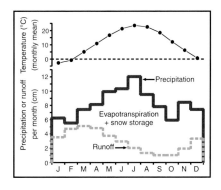

Scioto River at Columbus, Ohio (photo by K. Wilhelm).

Allegheny River

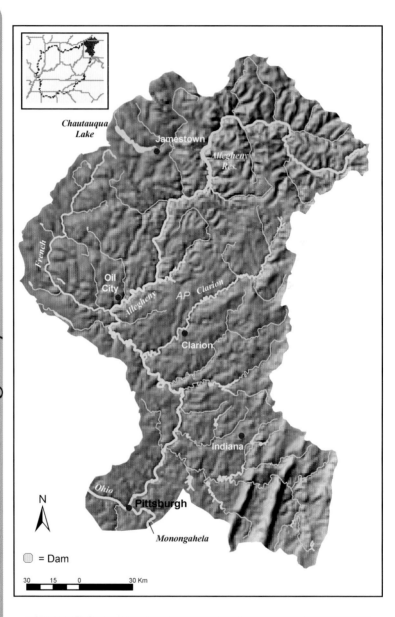

Relief: 690 m
Basin area: 30,300 km²
Mean discharge: 600 m³/s
Mean annual precipitation: 104 cm

Mean air temperature: 9°C
Mean water temperature: 11°C
No. of fish species: 114 to 120
No. of endangered species: 5

Physiographic province: Appalachian Plateaus (AP)

Major fishes: gizzard shad, common carp, bluegill, largemouth bass, spotted bass, channel catfish, walleye, sauger, emerald shiner, white crappie, gravel chub, blackchin shiner, river redhorse, black redhorse, longhead darter

Major other aquatic vertebrates: snapping turtle, stinkpot, midland painted turtle, spiny softshell turtle, common water snake, queen snake, bullfrog, green frog, hellbender, mudpuppy, osprey, great blue heron, mallard, beaver, muskrat

Major benthic insects: mayflies (*Hexagenia, Ephemerella, Isonychia*), stoneflies (*Allocapnia, Isoperla, Acroneuria*), caddisflies (*Hydroptila, Hydropsyche, Cheumatopsyche, Ceraclea*), true flies (*Tipula*)

Nonnative species: Asiatic clam, zebra mussel, common carp, goldfish, striped bass, yellow perch, brown trout, watercress, curly pondweed, brittle naiad, Eurasian watermilfoil, yellow iris, purple loosestrife

Major riparian plants: river birch, red maple, cottonwood, black willow, American sycamore, speckled alder

Special features: Upper Allegheny and

Clarion rivers designated as National Wild and Scenic Rivers; upper Allegheny has >100 undeveloped islands

Fragmentation: 8 navigation locks and dams on lower main stem, 12 major tributary dams

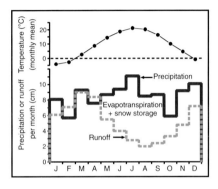

Allegheny River near New Kensington, Pennsylvania (photo by K. Wilhelm).

Monongahela River

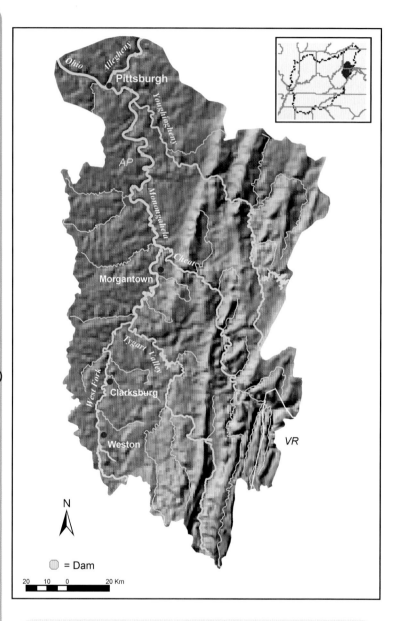

Relief: 1260 m
Basin area: 19,110 km²
Mean discharge: 377 m³/s
Mean annual precipitation: 106 cm

Mean air temperature: 10°C
Mean water temperature: 14°C
No. of fish species: >120
No. of endangered species: 0

Physiographic provinces: Valley and Ridge (VR), Appalachian Plateaus (AP)
Major fishes: gizzard shad, carp, bluegill, largemouth bass, spotted bass, channel catfish, walleye, sauger, emerald shiner, white crappie, black crappie, gravel chub, river redhorse, black redhorse, spotted sucker, flathead catfish, stoneroller, striped shiner, fantail darter, river carpsucker
Major other aquatic vertebrates: snapping turtle, stinkpot, midland painted turtle, spiny softshell turtle, common water snake, bullfrog, green frog, hellbender, belted kingfisher, osprey, bald eagle, coot, common loon, beaver, river otter, muskrat
Major benthic insects: mayflies (*Hexagenia*, *Brachycentrus*, *Ephemerella*, *Isonychia*, *Stenonema*), stoneflies (*Allocapnia*, *Isoperla*, *Acroneuria*), caddisflies (*Hydroptila*, *Hydropsyche*, *Cheumatopsyche*, *Ceraclea*), true flies (*Tipula*)
Nonnative species: Asiatic clam, zebra mussel, goldfish, carp, striped bass, alewife, tench, margined madtom, rainbow smelt, rainbow trout, brown trout, lake trout, yellow perch, purple loosestrife
Major riparian plants: river birch, red maple, cottonwood, black willow, American sycamore, speckled alder

Special features: headwaters of the Cheat and Tygart Valley rivers begin in the Monongahela National Forest
Fragmentation: 9 major navigation locks and dams, 5 major tributary dams

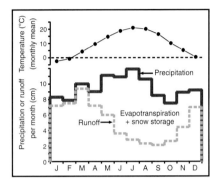

Monongahela River north of Morgantown, West Virginia (photo by Tim Palmer).

Chapter 10

Missouri River Basin

David L. Galat, Charles R. Berry, Jr., Edward J. Peters, and Robert G. White

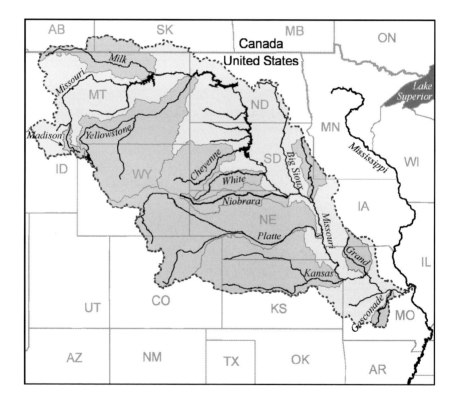

The Missouri River basin is the second largest in the United States, surpassed in area only by the Mississippi River basin of which it is a part. It drains about one-sixth of the conterminous United States ($1,371,017\,km^2$) and about $25,100\,km^2$ in Canada. It is the longest river in the United States ($4180\,km$) from its headwaters in Hell Roaring Creek, Montana, and the second longest river in North America following the Mackenzie-Slave-Peace rivers ($4241\,km$). The Missouri basin trends in a northwest to southeast direction across the north-central United States (see map) and includes all or parts of 10 states, 2 Canadian provinces, and 25 Native American Tribal Reservations or Lands ($74,500\,km^2$). Latitude of the basin ranges from $49.7°N$ in southwest Saskatchewan to $37.0°N$ in southwest Missouri. A wide range of climatic conditions, geological complexity, and topographic relief exist within the three physiographic divisions (Rocky Mountains, Great Plains, Central Lowlands) that contribute to the Missouri River basin. Most of the basin is semiarid with about half receiving $<41\,cm/yr$ precipitation, and 70% of this occurring as rainfall during the growing season. Although the Missouri's basin area is much greater than the combined area of the Upper Mississippi and Ohio rivers, its mean discharge of $1956\,m^3/s$ is much less than either. The Missouri's upper portion remains unchannelized for about $739\,km$, but it then becomes one of the most regulated rivers in the United States with a series of six major main stem impoundments.

About 20 Native American tribes belonging to four linguistic groups (Algonkian, Siouian, Caddoan, Shoshonean) lived in the Missouri River basin around 1500. Major tribal groups living along the river around 1800 from its mouth northwestward included the Oto, Missouri, Omaha, Ponca, Brule, Teton Sioux, Yankton Sioux, Yanktoinai Sioux, Arikara, Hidatsa (Minitari), Mandan, Assiniboin, Atsina, and Piegan Blackfoot. The Louisiana Purchase in 1803 put the entire Missouri River basin into Federal ownership and significant Euro-American expansion into the basin began in 1848. The Lewis and Clark expedition (1804–06) provided unprecedented detail of the unaltered Missouri's geography, natural history, and ethnography of its Native peoples. Following Lewis and Clark's "Corps of Discovery," the Missouri became the "original highway west" for Euro-American development of the United States. Today, private, county, state, or Native American tribes own about 86% of the basin.

Twelve rivers are described in this chapter, including the Missouri main stem. The Yellowstone, Platte, and Kansas rivers are the largest tributaries with basin areas well over $100,000\,km^2$ and means discharges $>200\,m^2/s$. There are many tributaries draining basin areas $>20,000\,km^2$ described in this chapter, however, including the White, Milk, Cheyenne, Big Sioux, Niobrara, and Grand rivers. Also described are the Madison and Gasconade rivers.

Missouri River

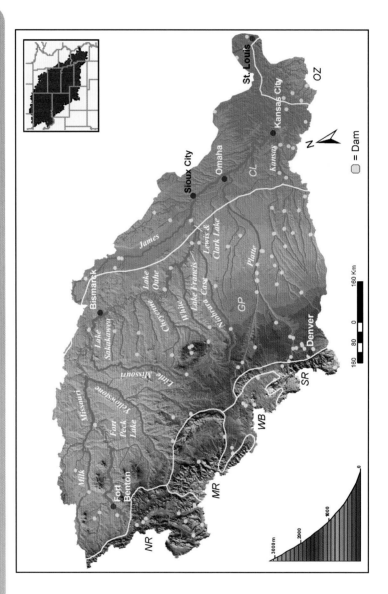

Relief: 4277 m
Basin area: 1,371,017 km^2
Mean discharge: 1956 m^3/s
Mean annual precipitation:
50 cm

Mean air temperature: 7.4°C
Mean water temperature: 9.3°C
No. of fish species: basin ~183
(138 native)
No. of endangered species: 7

Physiographic provinces: 7 provinces; most in Rocky Mountains (NR, MR, SR), Great Plains (GP), Central Lowland (CL)

Major fishes: shovelnose sturgeon, goldeye, gizzard shad, emerald shiner, red shiner, river shiner, flathead chub, spotfin shiner, river carpsucker, shorthead redhorse, white sucker, channel catfish, flathead catfish, white crappie, freshwater drum

Major other aquatic vertebrates: false map turtle, softshell turtles, great blue heron, wood duck, beaver

Major benthic insects: mayflies (*Baetis*, *Heptagenia*, *Ephemerella*, *Tricorythodes*, *Isonychia*, *Hexagenia*, *Caenis*, *Stenonema*), caddisflies (*Hydropsyche*, *Cheumatopsyche*, *Neureclipsis*, *Oecetis*)

Nonnative species: Russian olive, reed canary grass, Johnson grass, Asiatic clam, 28 fishes (rainbow trout, brown trout, cisco, common carp, goldfish, bighead carp)

Major riparian plants: cottonwoods, willows, American elm, green ash, box elder, red mulberry, Virginia creeper, prairie cordgrass, Canada wild rye, switchgrasses, giant ragweed, smartweeds

Special features: longest named river in North America (3768 km); primary water route for settlement of western United States

Fragmentation: 6 major main-stem dams; 581 total large dams

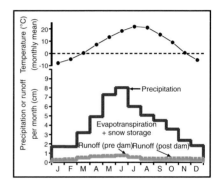

Missouri River in the Cow Island area, Montana (photo by C. Berry).

Yellowstone River

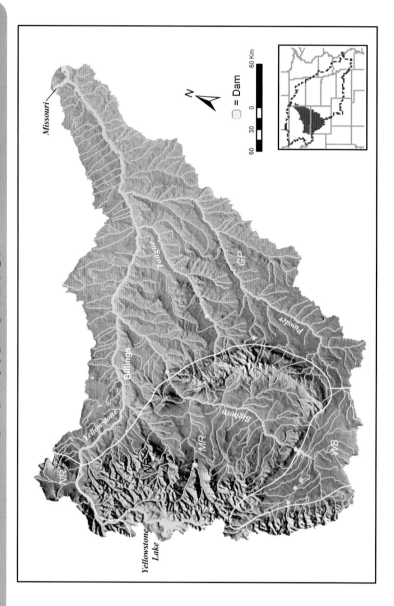

Relief: 3050 m
Basin area: 182,336 km^2
Mean discharge: 362 m^3/s
Mean annual precipitation: 30 cm

Mean air temperature: 7.8°C
Mean water temperature: NA
No. of fish species: 56 (36 native)
No. of endangered species: 1

Physiographic provinces: Northern Rocky Mountains (NR), Middle Rocky Mountains (MR), Wyoming Basin (WB), Great Plains (GP)

Major fishes: Yellowstone cutthroat trout, mountain whitefish, white sucker, longnose dace, mottled sculpin, shovelnose sturgeon, paddlefish, freshwater drum, channel catfish, sauger, goldeye, river carpsucker, shorthead redhorse, chubs, shiners

Major other aquatic vertebrates: river otter, beaver, bald eagle, osprey, great blue heron, spiny softshell turtle, snapping turtle, painted turtle

Major benthic insects: mayflies (*Baetis, Ephemerella, Epeorus, Ephemera, Ametropus, Lachlania, Ephoron, Caenis, Centroptilum, Isonychia*), stoneflies (*Acroneuria, Pteronarcys, Pteronarcella, Arcynopteryx, Paraleuctra, Capnia, Alloperla*), caddisflies (*Rhyacophila, Amiocentrus, Glossosoma,*

Brachycentrus, Lepidostoma, Neotrichia, Oecetis, Leptocella)

Nonnative species: black bullhead, black crappie, white crappie, brook trout, brown trout, rainbow trout, common carp, green sunfish, largemouth bass, northern pike, rainbow smelt, walleye

Major riparian plants: black cottonwood, narrowleaf cottonwood, plains cottonwood, Geyer willow, wolf willow

Special features: longest free-flowing river in conterminous U.S. (1091 km)

Fragmentation: no major dams

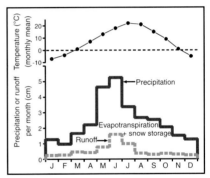

Yellowstone River, Montana (photo by Travel Montana).

White River

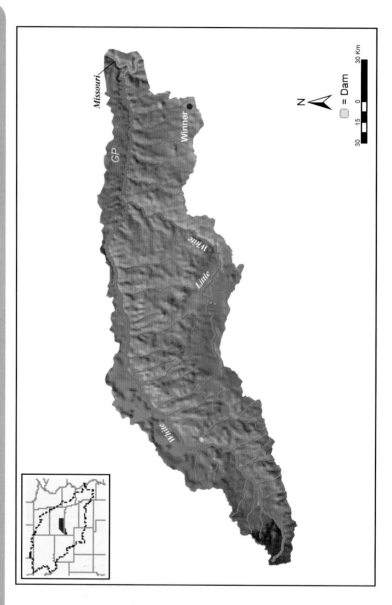

Relief: 1112 m
Basin area: 26,418 km²
Mean discharge: 16 m³/s
Mean annual precipitation: 44 cm

Mean air temperature: 9.3°C
Mean water temperature: 12.7°C
No. of fish species: 49 (41 native)
No. of endangered species: 1

Physiographic province: Great Plains (GP)

Major fishes: flathead chub, plains minnow, fathead minnow, sturgeon chub, common carp, sand shiner, western silvery minnow, channel catfish

Major other aquatic vertebrates: northern leopard frog, bullfrog, Great Plains toad, Woodhouse's toad, chorus frog, tiger salamander, common snapping turtle

Major benthic insects: NA

Nonnative species: Canada thistle, saltcedar, Russian olive, smooth brome, leafy spurge, brown trout, rainbow trout, brook trout, common carp, black crappie, white crappie, largemouth bass, bluegill

Major riparian plants: plains cottonwood, green ash, sandbar willow, buffalo grass

Special features: harsh aquatic conditions; badlands and xeric landscape; Little White River (major tributary) has more benign conditions

Fragmentation: no main-stem dams

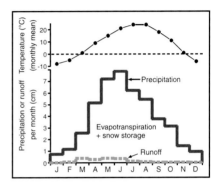

White River, South Dakota, near border of Pine Ridge Indian Reservation and Badlands National Park (photo by C. Berry).

Platte River

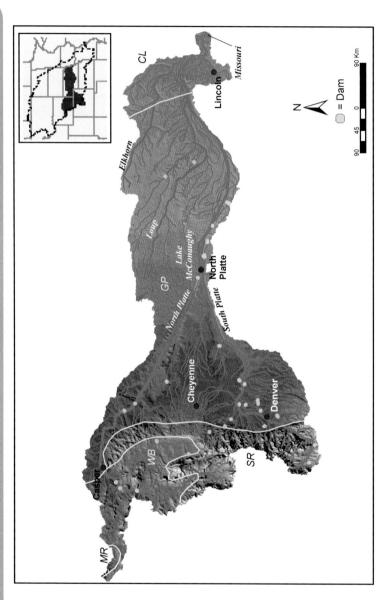

Relief: 3524 m
Basin area: 230,362 km^2
Mean discharge: 203 m^3/s
Mean annual precipitation: 50 cm

Mean air temperature: 9.1°C
Mean water temperature: 11.8°C
No. of fish species: 100 (76 native)
No. of endangered species: 17

Physiographic provinces: Southern Rocky Mountains (SR), Wyoming Basin (WB), Great Plains (GP), Central Lowland (CL)

Major fishes: shovelnose sturgeon, longnose gar, flathead chub, speckled chub, sand shiner, red shiner, river shiner, western silvery minnow, plains minnow, river carpsucker, quillback, plains killifish, channel catfish, freshwater drum

Major other aquatic vertebrates: snapping turtle, spiny softshell turtle, painted turtle, migratory waterfowl, beaver, muskrat

Major benthic insects: mayflies (*Heptagenia, Hexagenia, Caenis, Baetis, Isonychia*), stoneflies (*Acroneuria, Isoperla, Pteronarcys*), caddisflies (*Hydropsyche, Cheumatopsyche*), true flies (*Chaoborus, Simulium, Chernovskiia, Chironomus, Robackia, Saetheria*)

Nonnative species: 24 fishes (brown trout, rainbow trout, common carp, grass carp, bighead carp, western mosquitofish, striped bass, yellow perch, walleye)

Major riparian plants: eastern cottonwood, eastern red cedar, rough leaf dogwood, silver maple, green ash, sandbar willow

Special features: a wide, shallow braided river with shifting sandbars

Fragmentation: 7 main-stem dams, 20 diversions

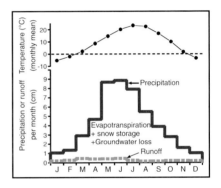

Platte River near North Bend, Nebraska (photo by E. Peters).

Gasconade River

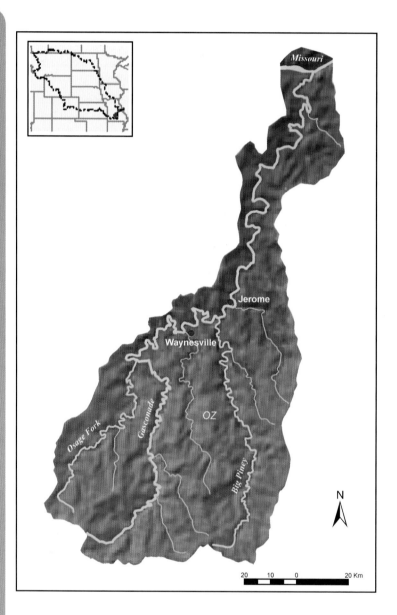

Relief: 380 m
Basin area: 9258 km²
Mean discharge: 87 m³/s
Mean annual precipitation: 108 cm

Mean air temperature: 13.4°C
Mean water temperature: 15.2°C
No. of fish species: 105 (98 native)
No. of endangered species: 4

Physiographic province: Ozark Plateaus (OZ)

Major fishes: bleeding shiner, Ozark minnow, largescale stoneroller, wedgespot shiner, striped shiner, bigeye shiner, southern redbelly dace, longear sunfish, smallmouth bass, rock bass, northern orangethroat

Major other aquatic vertebrates: bullfrog, green frog, common map turtle, red-eared slider, midland smooth softshell turtle, common snapping turtle, western painted turtle, northern water snake, wood duck, belted kingfisher, beaver, mink, muskrat, river otter

Major benthic insects: mayflies (*Acentrella*, *Baetis*, *Isonychia*, *Stenonema*), caddisflies (*Chimarra*, *Cheumatopsyche*, *Hydropsyche*), stoneflies (*Taeniopteryx*, *Strophopterx*, *Neoperla*)

Nonnative species: Asiatic clam, rainbow trout, common carp, goldfish

Major riparian plants: silver maple, American sycamore, green ash, river birch, Ward's willow, swamp dogwood, buttonbush, water willow

Special features: located in karst topography, sections lose flow to groundwater, 76 springs; largest undammed tributary to Missouri River draining Ozark Plateau; bluestripe and Missouri saddled darters endemic to Gasconade and adjacent drainages in Missouri

Fragmentation: no main-stem dams

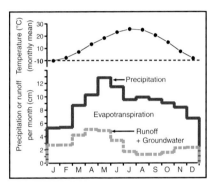

Gasconade River, Missouri (poto by J. Rathert, Missouri Department of Conservation).

Madison River

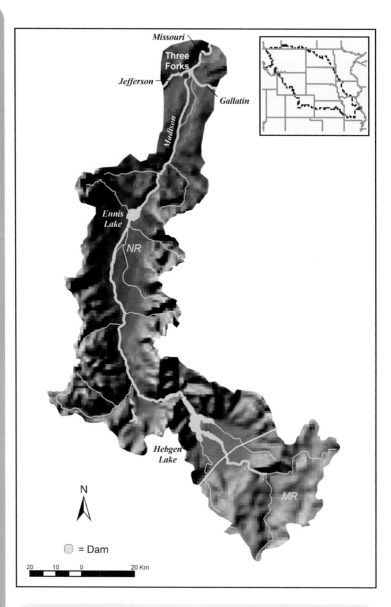

Missouri

Three Forks

Jefferson

Gallatin

Madison

Ennis Lake

NR

Hebgen Lake

MR

N

◯ = Dam

20   10   0   20 Km

Relief: 1959 m
Basin area: 6537 km^2
Mean discharge: 50.7 m^3/s
Mean annual precipitation: 34 cm

Mean air temperature: 6.4°C
Mean water temperature: NA
No. of fish species: 17
(10 native)
No. of endangered species: 0

Physiographic province: Northern Rocky Mountains (NR), Middle Rocky Mountains (MR)

Major fishes: westslope cutthroat trout, rainbow trout, brown trout, mountain whitefish, mountain sucker, longnose sucker, white sucker, longnose dace, mottled sculpin

Major other aquatic vertebrates: northern leopard frog, spotted frog, osprey, bald eagle, American dipper, great blue heron, common merganser, river otter, beaver, water shrew

Major benthic insects: mayflies (*Baetis, Epeorus, Tricorythodes, Rhithrogena, Paraleptophlebia, Ephemera*), caddisflies (*Cheumatopsyche, Hydropsyche, Brachycentrus, Micrasema, Glossosoma*), stoneflies (*Pteronarcys, Acroneuria, Claassenia, Isoperla*)

Nonnative species: rainbow trout, brown trout, brook trout, Utah chub,

New Zealand mud snail, salmonid whirling disease parasite

Major riparian plants: black cottonwood, narrowleaf cottonwood, Geyer willow, wolf willow, sandbar willow

Special features: geothermal inputs, notably near headwaters in geyser basins of Yellowstone National Park

Fragmentation: Madson (Ennis), Hegben, and Quake Lake dams on main stem

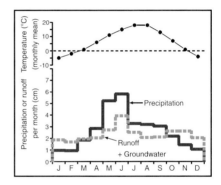

Madison River near West Yellowstone, Wyoming (photo by C. E. Cushing).

Milk River

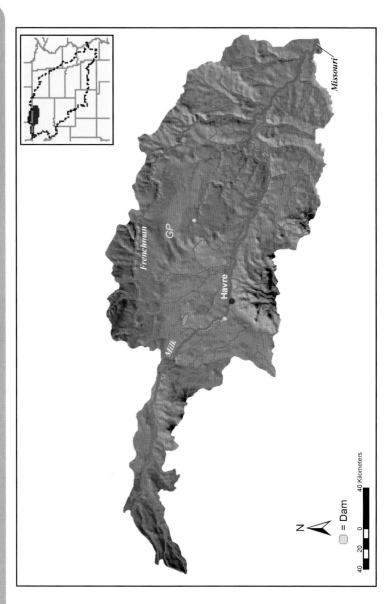

Relief: 1930 m
Basin area: 57,839 km²
Mean discharge: 18.9 m³/s
Mean annual precipitation: 29 cm

Mean air temperature: 6.1°C
Mean water temperature: NA
No. of fish species: 45 (30 native)
No. of endangered species: 0

Physiographic province: Great Plains (GP)

Major fishes: shovelnose sturgeon, paddlefish, freshwater drum, channel catfish, sauger, Iowa darter, river carpsucker, blue sucker, bigmouth buffalo, smallmouth buffalo, shorthead redhorse, brook stickleback, chubs, shiners

Major other aquatic vertebrates: osprey, common merganser, snapping turtle, spiny softshell turtle, painted turtle, muskrat

Major benthic insects: mayflies (*Analetris, Camelobaetidius, Ametropus, Lachlania, Raptoheptagenia, Macdunnoa, Cercobrachys, Hexagenia*), stoneflies (*Oemopteryx, Acroneuria*)

Nonnative species: common carp, spottail shiner, bluegill, smallmouth bass, white crappie, black crappie, yellow perch, walleye, black bullhead, rainbow trout, brown trout, lake whitefish, northern pike, virile crayfish

Major riparian plants: plains cottonwood, red-osier dogwood, peach-leaved willow, Rocky Mountain juniper, box elder

Special features: meandering, highly braided channel with very unstable sand bottom in upper reaches; reach above Fresno Dam one of few remnant Great Plains river ecosystems

Fragmentation: Fresno dam on main stem; one municipal water weir; 4 diversion dams

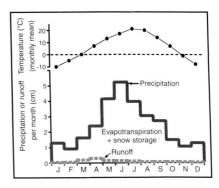

Milk River east of Havre, Montana (photo by Tim Palmer).

Cheyenne River

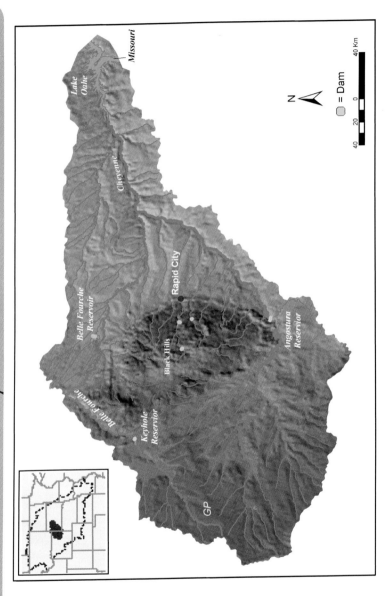

Relief: 1410 m
Basin area: 63,455 km^2
Mean discharge: 25 m^3/s
Mean annual precipitation: 51 cm

Mean air temperature: 7.6°C
Mean water temperature: 12.7°C
No. of fish species: 56 (37 native)
No. of endangered species: 2

Physiographic province: Great Plains (GP)

Major fishes: brook trout, brown trout, rainbow trout, longnose dace, white sucker, green sunfish, sand shiner, fathead minnow, common carp, flathead chub, channel catfish, plains minnow, shorthead redhorse, red shiner

Major other aquatic vertebrates: northern leopard frog, Great Plains toad, Woodhouse's toad, chorus frog, plains spadefoot toad, tiger salamander, common snapping turtle, spiny softshell turtle

Major benthic insects: caddisflies (*Glossosoma*, *Hydropsyche*, *Hesperophylax*), mayflies (*Baetis*), stoneflies (*Acroneuria*, *Isoperla*), true flies (*Tipula*, *Atherix*)

Nonnative species: tamarisk, Russian olive, Canada thistle, leafy spurge, smooth brome, common carp, brown trout, rainbow trout, brook trout, cutthroat trout, largemouth bass, bluegill, pumpkinseed, smallmouth bass, rock bass, golden shiner

Major riparian plants: white spruce, aspen, Bebb willow, plains cottonwood, green ash, salt grass, buffalo grass, sandbar willow

Special features: Black Hills coldwater streams and hot springs; harsh conditions; section proposed for National Wild and Scenic River

Fragmentation: three major dams on main stem; Lake Oahe isolates Cheyenne from other Missouri River tributaries

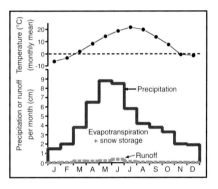

Cheyenne River near Oral, South Dakota, below Angostura dam (photo by C. Berry).

Big Sioux River

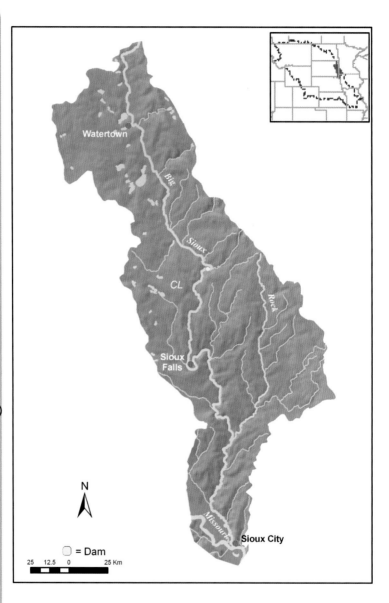

N

= Dam

25   12.5   0   25 Km

Relief: 305 m   Mean air temperature: 7.8°C
Basin area: 23,325 km²   Mean water temperature: 12.0°C
Mean discharge: 35.4 m³/s   No. of fish species: 70 (66 native)
Mean annual precipitation: 62 cm   No. of endangered species: 2

Physiographic province: Central Lowland (CL)

Major fishes: northern pike, common carp, common shiner, fathead minnow, red shiner, sand shiner, river carpsucker, shorthead redhorse, white sucker, black bullhead, channel catfish, tadpole madtom, brook stickleback, green sunfish, walleye, Johnny darter, freshwater drum

Major other aquatic vertebrates: American toad, Canadian toad, chorus frog, Great Plains toad, northern leopard frog, tiger salamander, spiny softshell turtle, common snapping turtle, western painted turtle, beaver, mink

Major benthic insects: caddisflies (*Cheumatopsyche*, *Hydroptila*), mayflies (*Baetis intercalaris*, *Hexagenia*, *Caenis*, *Stenacron*), true flies (*Chironomus*)

Nonnative species: bighead carp, common carp, grass carp, smallmouth bass, smooth brome grass, reed canary grass, Russian olive, buckthorn, Canada thistle, bindweed

Major riparian plants: green ash, American elm, eastern cottonwood, sandbar willow, smooth brome grass, reed canary grass

Special features: relatively intact floodplain wetlands and river channel; geological features related to quartzite formations; recreational use for hunting and fishing

Fragmentation: Sioux Falls is natural barrier to fish movement; three low-head (2 to 5 m high) dams impede fish movements

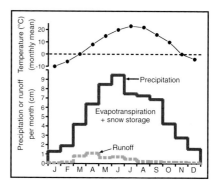

Big Sioux River at Sioux Falls, South Dakota (photo by Tim Palmer).

Niobrara River

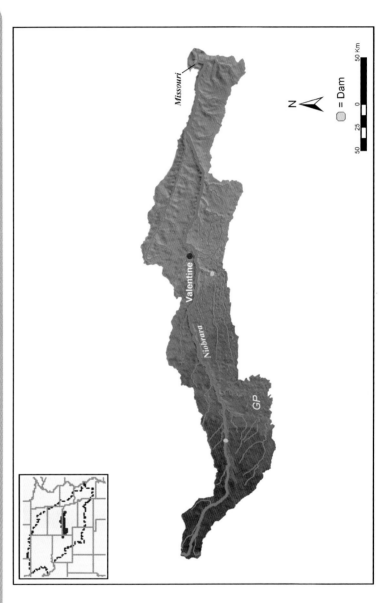

Relief: 1182 m
Basin area: 32,600 km²
Mean discharge: 49 m³/s
Mean annual precipitation: 47 cm

Mean air temperature: 8.3°C
Mean water temperature: 12.3°C
No. of fish species: 67 (43 native)
No. of endangered species: 6

Physiographic province: Great Plains (GP)

Major fishes: sand shiner, red shiner, river shiner, emerald shiner, bigmouth shiner, flathead chub, river carpsucker, channel catfish

Major other aquatic vertebrates: spiny softshell turtle, painted turtle, beaver, muskrat, river otter

Major benthic insects: mayflies (*Isonychia, Caenis, Baetis, Pseudocleon, Callibaetis*), caddisflies (*Hydropsyche, Brachycentrus*), true flies (*Rheocricotopus*)

Nonnative species: 24 fishes (brown trout, rainbow trout, brook trout, alewife, common carp)

Major riparian plants: eastern red cedar, ponderosa pine, eastern

cottonwood, box elder, paper birch, American elm

Special features: swift-flowing Great Plains prairie river with eastern, western, and northern forest species in riparian zone; groundwater a major source of discharge

Fragmentation: 3 main-stem dams

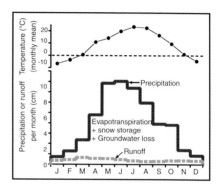

Niobrara River near Merriman, Nebraska (photo by Tim Palmer).

Kansas River

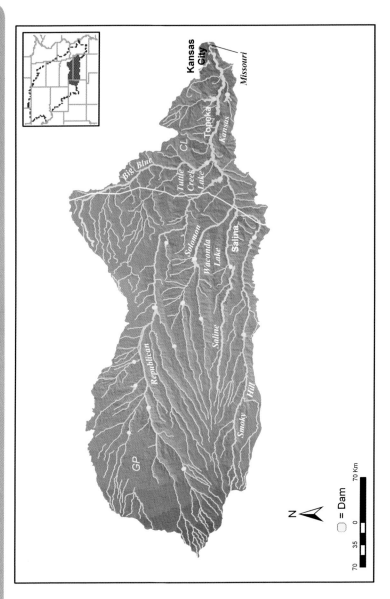

Relief: 975 m
Basin area: 159,171 km²
Mean discharge: 214 m³/s
Mean annual precipitation: 61 cm

Mean air temperature: 12.2°C
Mean water temperature: 11.6°C
No. of fish species: 99 (75 native)
No. of endangered species: 1

Physiographic provinces: Great Plains (GP), Central Lowland (CL)

Major fishes: shovelnose sturgeon, longnose gar, gizzard shad, creek chub, suckermouth minnow, plains minnow, sand shiner, red shiner, river carpsucker, shorthead redhorse, blue sucker, white sucker, flathead catfish, channel catfish, largemouth bass, sauger, freshwater drum

Major other aquatic vertebrates: smooth softshell turtle, migratory waterfowl, beaver, muskrat

Major benthic insects: mayflies (*Isonychia*, *Heptagenia*, *Tricorythodes*), stoneflies (*Neoperla*, *Isoperla*), caddisflies (*Hydropsyche*, *Cheumatopsyche*)

Nonnative species: 24 fishes (common carp, grass carp, bighead carp, western mosquitofish, striped bass, yellow perch, walleye)

Major riparian plants: eastern red cedar, eastern cottonwood, box elder, American elm, sycamore, silver maple, willows

Special features: a large river that starts in the Great Plains and ends in the Central Lowland

Fragmentation: 18 large reservoirs, >13,000 small impoundments, dewatering by irrigation withdrawals

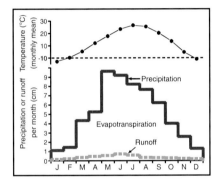

Kansas River in vicinity of Lawrence, Kansas (photo by Tim Palmer).

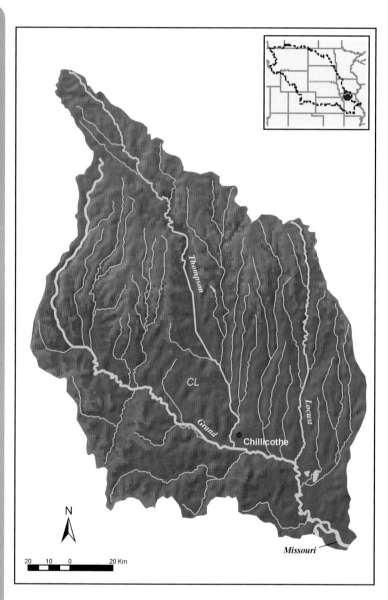

Grand River

Relief: 238 m
Basin area: 20,390 km²
Mean discharge: 117 m³/s
Mean annual precipitation: 92 cm

Mean air temperature: 11.8°C
Mean water temperature: 13.0°C
No. of fish species: 61 (55 native)
No. of endangered species: 3

Physiographic province: Central Lowland (CL)

Major fishes: shortnose gar, bigmouth shiner, red shiner, creek chub, sand shiner, central stoneroller, fathead minnow, bluntnose minnow, common carp, river carpsucker, channel catfish, flathead catfish, bluegill, green sunfish, Johnny darter

Major other aquatic vertebrates: bullfrog, green frog, false map turtle, red-eared slider, smooth softshell turtle, common snapping turtle, western painted turtle, northern water snake, wood duck, belted kingfisher, beaver, mink, muskrat, river otter

Major benthic insects: mayflies (*Baetis, Isonychia, Stenonema, Caenis*), stoneflies (*Neoperla*), caddisflies (*Cheumatopsyche, Hydropsyche, Potamyia*)

Nonnative species: common carp, goldfish, bighead carp, grass carp, rainbow smelt, white perch, striped bass, Asiatic clam, Johnson grass

Major riparian plants: cottonwood, silver maple, green ash, American sycamore, hackberry, aromatic sumac, gray dogwood, giant ragweed, stinging nettle, poison ivy, grapes

Special features: preglacial channel of ancestral Missouri River; largest prairie river in Missouri relatively unaffected by dams or channelization (mouth to Rkm 56)

Fragmentation: no main-stem dams; ~30 impoundments >20 ha on tributaries

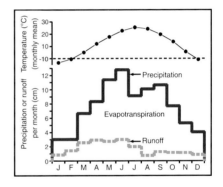

Grand River at Holmes Bend access in Daviess County, Missouri (photo by G. Pitchford).

Chapter 11

Colorado River Basin

Dean W. Blinn and N. LeRoy Poff

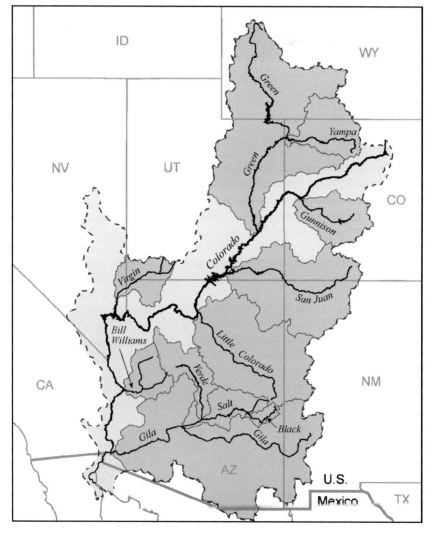

The Colorado River Basin lies within the Intermontane Plateaus of the American Southwest. The basin encompasses an area from 42° to 32°N latitude ranging from the high mountains of the Rockies in Wyoming and Colorado to the Colorado River delta in Mexico (see map). The Colorado basin drains seven states, or nearly 8% of the United States, including parts of Colorado, Wyoming, Utah, New Mexico, Nevada, California, and 95% of Arizona. The entire basin encompasses 642,000 km^2 with the Upper Basin draining about 45% of the total. The majority (75%) of the river's flow is supplied by mountain headwater streams, but most of the catchment lies in a semi-arid desert. Therefore, the low runoff (~3 cm/yr) makes the Colorado basin one of the driest in the world. In spite of this water scarcity, there are heavy demands from urbanization, agriculture, and hydropower. Nearly 64% of the runoff is used for irrigation and another 32% is lost by evaporation from reservoirs[1] resulting in less than 10% (40 m^3/s) of the Colorado's virgin discharge (550 m^3/s) reaching the Gulf of California.

The Upper Basin shows evidence of widespread human culture dating 11,000 years ago.[2] About two millennia ago, the agrarian Anasazi culture arose in the southern part of the basin as seen in numerous cliff dwellings, including Mesa Verde National Park. Around 800 A.D., the Fremont Culture arose throughout most of Utah and western Colorado, but they were supplanted by the Utes, who ultimately spread to the Upper Basin by the arrival of the Spanish in the 1630s. The great influx of English-speaking settlers into the Upper Basin began during the mid 1800s. The Lower Basin also has a long history of agricultural occupation including the "Mogollon culture" (as early as 200 B.C.), and the Hohokam culture. Spanish explorers ventured into the Lower Basin from Mexico in the late 1600s and Mormon settlers moved into the Little Colorado basin in the late 1800s.

Twenty-two major rivers converge with the Colorado after it begins its descent from Rocky Mountain National Park and winds through the plateaus of Colorado, Utah, and Arizona, onto the deserts of southwestern Arizona, and finally into the Gulf of California. This chapter describes 11 of these tributaries in addition to the main stem. The major flow to the Colorado River comes from tributaries of the Upper Basin, particularly the Green (172 m^3/s), Yampa (64 m^3/s), Gunnison (74 m^3/s), and San Juan rivers (65 m^3/s). Other rivers cover large areas, but contribute relatively little flow (Little Colorado, Gila). Also described are the Virgin, Bill Williams, Black, Verde, and Salt rivers.

1 Dynesius, M., & Nilsson, C. 1994. *Fragmentation and flow regulation of river systems in the northern third of the world.* Science 266:753–762.

2 Smith, A. M. 1974. Ethnography of the Northern Utes. Papers in Anthropology no. 17. 1974. Museum of New Mexico Press, Albuquerque, New Mexico.

Colorado River

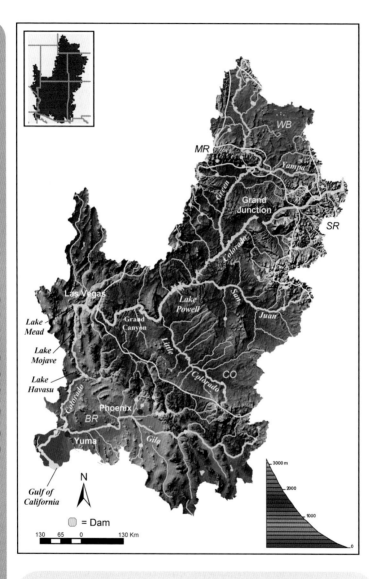

Relief: ~4100 m
Basin area: 642,000 km²
Mean discharge: 550 m³/s
(virgin) ~40 m³/s (present)
Mean annual precipitation: 22 cm
(upper basin), 11 cm (lower
basin)

Mean air temperature: 14°C
(upper basin), 21.1°C (lower
basin)
Mean water temperature: 11°C
(upper basin), 21°C (lower basin)
No. of fish species: ~75 to 85
(42 native)
No. of endangered species: 16

Physiographic provinces: Middle Rocky Mountains (MR), Southern Rocky Mountains (SR), Wyoming Basin (WB), Colorado Plateaus (CO), Basin and Range (BR), Baja California (BC)

Major fishes: bonytail chub, brook trout, brown trout, channel catfish, common carp, fathead minnow, flannelmouth sucker, humpback chub, longfin dace, rainbow trout, razorback sucker, red shiner, roundtail chub, speckled dace

Major other aquatic vertebrates: American coot, American widgeon, Arizona toad, beaver, bullfrog, great blue heron, mallard, muskrat, snowy egret, Sonoran mud turtle

Major benthic insects: mayflies (*Baetis*, *Drunella*, *Ephemerella*, *Epeorus*, *Heptagenia*, *Paraleptophlebia*, *Traverella*, *Tricorythodes*), stoneflies (*Capnia*, *Paraleuctra*, *Suwallia*), caddisflies (*Ceratopsyche*, *Chimarra*, *Hydropsyche*, *Hydroptila*, *Ochrotrichia*, *Oecetis*)

Nonnative species: Asiatic clam, bullfrog, >30 fishes (carp, channel catfish, rainbow trout, threadfin shad), New Zealand mudsnail, red swamp crayfish, saltcedar, virile crayfish, water fern

Major riparian plants: Arizona sycamore, bulrush, coyote willow, Fremont cottonwood, gamble oak, narrowleaf cottonwood, saltcedar

Special features: drains nearly 8% of U.S.; runs through Grand Canyon

Fragmentation: over 40 flow-regulation structures, 4 large main-stem reservoirs; one of the most regulated rivers in the world

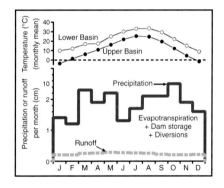

Colorado River, Grand Canyon, Arizona.

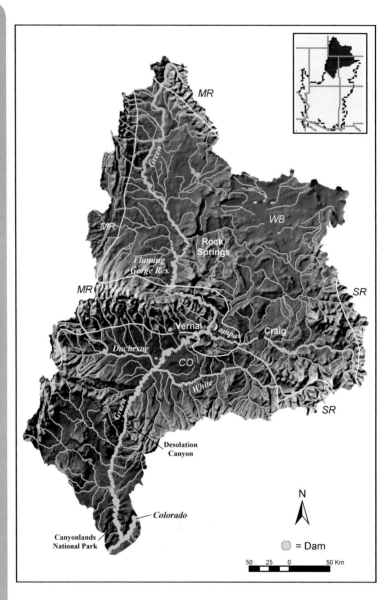

Green River

Relief: ~2950 m
Basin area: 116,200 km²
Mean discharge: 172 m³/s
Mean annual precipitation: 32 cm

Mean air temperature: 6.3°C
Mean water temperature: 2°C to
25°C at Rkm 189
No. of fish species: 37 (12 native)
No. of endangered species: 3

Physiographic provinces: Middle Rocky Mountains (MR), Wyoming Basin (WB), Southern Rocky Mountains (SR), Colorado Plateau (CO)
Major fishes: bonytail chub, Colorado River cutthroat trout, Colorado pikeminnow, humpback chub, razorback sucker, roundtail chub, rainbow trout
Major other aquatic vertebrates: beaver, boreal western toad, Clark's grebe, muskrat, northern leopard frog, spadefoot toad, tiger salamander, white-faced ibis, whooping crane
Major benthic insects: mayflies (*Baetis*, *Drunella*, *Ephemerella*, *Heptagenia*), stoneflies (*Arcynopteryx*, *Capnia*, *Hesperoperla*, *Taenionema*), caddisflies (*Brachycentrus*, *Helicopsyche*, *Hesperophylax*, *Hydroptila*, *Leucotrichia*, *Oecetis*, *Psychoglypha*, *Rhyacophila*)
Nonnative species: channel catfish, common carp, fathead minnow, rainbow trout, saltcedar, signal crayfish, virile crayfish

Major riparian plants: alkali sacaton, big sagebush, black greasewood, Fremont cottonwood, saltcedar, saltgrass
Special features: large desert river; important habitat for endangered fishes indigenous to Colorado River system
Fragmentation: major reservoirs on main stem in upper basin (Flaming Gorge and Fontenelle); major tributaries extensively dammed or diverted, except Yampa River

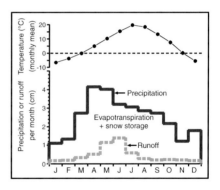

Green River in Browns Park National Wildlife Refuge, northeastern Utah (photo by D. M. Merritt).

Yampa River

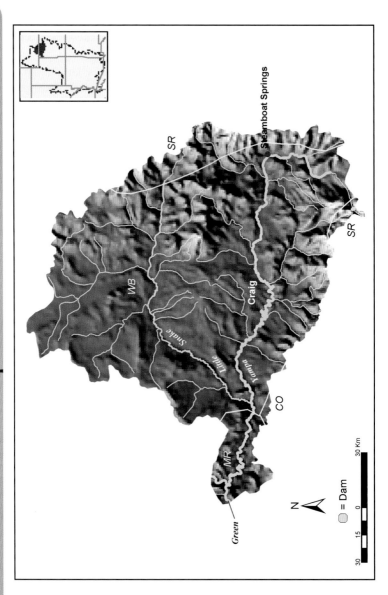

Relief: ~2276 m

Basin area: 24,595 km²

Mean discharge: 64 m³/s

Mean annual precipitation: 43 cm

Mean air temperature: 5.9°C

Mean water temperature: 9°C

No. of fish species: 30 (12 native)

No. of endangered species: 4

Physiographic provinces: Southern Rocky Mountains (SR), Middle Rocky Mountains (MR), Wyoming Basin (WB)

Major fishes: bluehead sucker, bonytail chub, channel catfish, Colorado pikeminnow, Colorado River cutthroat trout, flannelmouth sucker, green sunfish, humpback chub, mountain sucker, mountain whitefish, northern pike, razorback sucker, roundtail chub, speckled dace

Major other aquatic vertebrates: beaver, boreal western toad, Great Basin spadefoot toad, muskrat, northern leopard frog, river otter, wood frog

Major benthic insects: mayflies (*Baetis, Choroterpes, Ephemerella, Heptagenia, Paraleptophlebia, Tricorythodes, Rhithrogena, Traverella*), stoneflies (*Alloperla, Isoperla, Pteronarcella, Pteronarcys*), caddisflies (*Cheumatopsyche, Helicopsyche, Hydropsyche, Lepidostoma, Oecetis, Polycentropus*)

Nonnative species: channel catfish, green sunfish, northern pike, saltcedar, virile crayfish

Major riparian plants: biennial sage, Canada bluegrass, coyote willow, foxtail, Fremont cottonwood, horseweed, slender wheatgrass, western wheatgrass

Special features: last remaining free-flowing river in upper Colorado Basin; montane, high plains, and canyon river with perennial flow; critical to recovery of endangered native fishes in upper basin

Fragmentation: minimal, small headwater dams and water abstraction ≤10% of annual flow

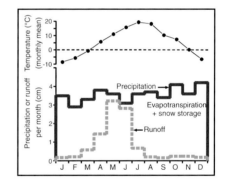

Yampa River, Colorado (photo by D. J. Cooper).

Little Colorado River

Relief: ~2600 m

Basin area: ~69,000 km²

Mean discharge: 6.5 m³/s

Mean annual precipitation: 17 cm

Mean air temperature: 13.5°C

Mean water temperature: 18°C

No. of fish species: 33 (9 native)

No. of endangered species: 3

Physiographic province: Colorado Plateaus (CO)

Major fishes: Apache trout, bluehead sucker, brown trout, fathead minnow, flannelmouth sucker, green sunfish, humpback chub, Little Colorado River sucker, Little Colorado spinedace, rainbow trout, roundtail chub, speckled dace

Major other aquatic vertebrates: Arizona toad, beaver, bullfrog, Chiricahua leopard frog, great blue heron, belted kingfisher, mallard, muskrat, red-spotted toad, Sonoran mud turtle, striped chorus frog

Major benthic insects: mayflies (*Baetis, Cinygmula, Drunella, Epeorus, Tricorythodes*), caddisflies (*Atopsyche, Cheumatopsyche, Glossosoma, Helicopsyche, Hesperophylax, Hydropsyche, Limnephilus, Oecetis, Oligophlebodes, Onocosmoecus, Polycentropus, Rhyacophila*), stoneflies (*Claassenia, Sweltsa*)

Nonnative species: Asiatic clam, brown trout, bullfrog, rainbow trout, fathead minnow, green sunfish, saltcedar, virile crayfish

Major riparian plants: Bebb willow, bulrush, cattails, common reed, Goodding willow, thin-leaf alder, saltcedar

Special features: ephemeral desert river; no flow >60% of year; headwaters recommended for conservation by World Wildlife Fund

Fragmentation: no major dams but numerous diversions

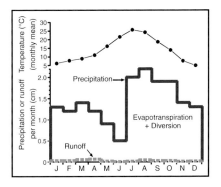

Little Colorado River near Holbrook, Arizona (photo by D. Blinn).

Gila River

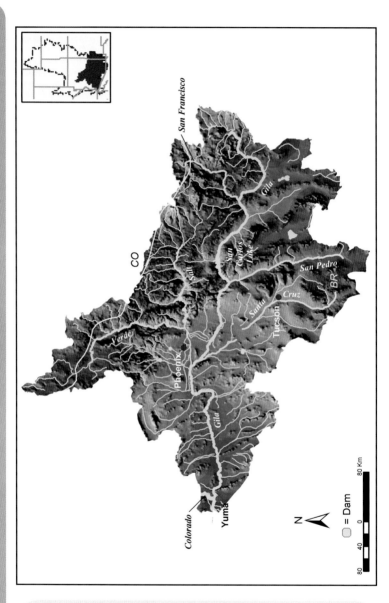

Relief: ~3050 m
Basin area: ~149,832 km^2
Mean discharge: >40 m^3/s (virgin), <6 m^3/s present
Mean annual precipitation: 25 cm

Mean air temperature: 15.2°C
Mean water temperature: 21°C
No. of fish species: 36 (19 native)
No. of endangered species: 2

Physiographic provinces: Basin and Range (BR), Colorado Plateaus (CO)

Major fishes: channel catfish, desert sucker, flathead catfish, loach minnow, longfin dace, roundtail chub, smallmouth bass, Sonora sucker, speckled dace, spike dace

Major other aquatic vertebrates: beaver, bullfrog, muskrat, Arizona toad, Chiricahua leopard frog, lowland leopard frog, Sonoran mud turtle, red-spotted toad, Sonoran Desert toad, Woodhouse toad

Major benthic insects: mayflies (*Choroterpes, Epeorus, Ephemerella, Serratella, Thraulodes, Tricorythodes, Traverella*), caddisflies (*Cheumatopsyche, Chimarra, Helicopsyche, Hydropsyche, Polycentropus, Oecetis, Protoptila, Ochrotrichia, Smicridea, Zumatrichia*), true flies (chironomid midges, black flies)

Nonnative species: bullfrog, channel catfish, *Daphnia lumholtzi*, flathead catfish, Rio Grande leopard frog, saltcedar, smallmouth bass, virile crayfish

Major riparian plants: Arizona sycamore, Bebb willow, box elder, Fremont cottonwood, Goodding willow, green ash, mesquite, narrowleaf cottonwood, thin-leaf alder, saltcedar

Special features: several sites in upper basin recommended for conservation by the World Wildlife Fund; ≥800 bird species in Gila Cliff Valley

Fragmentation: Coolidge, Painted Rock, and Gillespie dams on main stem; numerous main-stem diversions

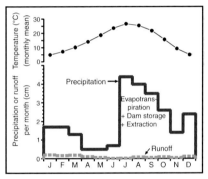

Upper Gila River at the Gila Preserve, Grant County, New Mexico (photo by D. M. Merritt).

Gunnison River

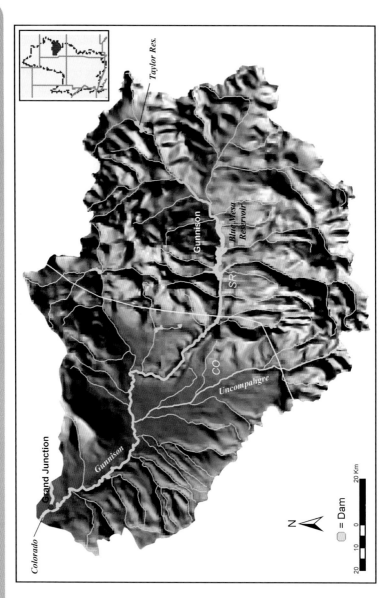

Relief: ~2600 m
Basin area: 21,000 km²
Mean discharge: 74 m³/s
Mean annual precipitation: 27 cm

Mean air temperature: 9.5°C
Mean water temperature: 10°C
No. of fish species: 30
(11 native) and 4 hybrids
No. of endangered species: 4

Physiographic provinces: Southern Rocky Mountains (SR), Colorado Plateaus (CO)

Major fishes: bluehead sucker, bonytail chub, Colorado pikeminnow, Colorado River cutthroat trout, flannelmouth sucker, humpback chub, mottled sculpin, mountain whitefish, razorback sucker, roundtail chub, speckled dace

Major other aquatic vertebrates: beaver, boreal western toad, Great Basin spadefoot toad, muskrat

Major benthic insects: mayflies (*Baetis*, *Heptagenia*, *Ephemerella*), stoneflies (*Isoperla*), caddisflies (*Hydropsyche*), true flies (*Atherix*, chironomid midges, black flies)

Nonnative species: brown trout, common carp, rainbow trout, red shiner, sand shiner, virile crayfish, white sucker

Major riparian plants: bent grass, bluegrass, box elder, Canada bluegrass, canary grass, goldenrod, muhly, scouring rush, smooth horsetail, spikerush, woolly sedge

Special features: critical habitat for 4 endangered Colorado River fishes; river runs through Black Canyon of Gunnison National Park

Fragmentation: highly fragmented by headwater

dams and 3 main-stem dams of Aspinall Unit; transbasin diversion through Gunnison Tunnel in lower river

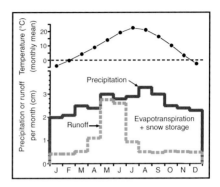

Gunnison River above Whitewater, Colorado (photo by Tim Palmer).

San Juan River

Relief: ~2800 m
Basin area: 59,600 km²
Mean discharge: 65 m³/s
Mean annual precipitation:
25 cm

Mean air temperature: 11.3°C
Mean water temperature: 12°C
No. of fish species: 26
(7 native) and 3 hybrids
No. of endangered species: 2

Physiographic provinces: Southern Rocky Mountains (SR), Colorado Plateaus (CO)

Major fishes: bluehead sucker, channel catfish, Colorado pikeminnow, common carp, fathead minnow, flannelmouth sucker, mottled sculpin, razorback sucker, roundtail chub, speckled dace

Major other aquatic vertebrates: beaver, boreal western toad, bullfrog, muskrat, northern leopard frog, northern water shrew, plains leopard frog, tiger salamander, western chorus frog

Major benthic insects: mayflies (*Acentrella*, *Baetis*, *Callibaetis*, *Drunella*, *Epeorus*, *Ephemera*, *Ephemerella*, *Heptagenia*, *Rhithrogena*, *Tricorythodes*), caddisflies (*Cheumatopsyche*, *Hydropsyche*, *Hydroptila*, *Oecetis*, *Ochrotrichia*, *Smicridea*), stoneflies (*Isoperla*)

Nonnative species: black bullhead, channel catfish, common carp, red shiner, fathead minnow, largemouth bass, Rio Grande killifish, Russian olive, saltcedar

Major riparian plants: Fremont cottonwood, Russian olive, upland herbs and shrubs, wetland herbs, willows

Special features: large desert river; important for the recovery of 2 native endangered fishes; mimicking of natural flow regime

implemented for restoration; runs through Four Corners

Fragmentation: Navajo Reservoir in upper basin; flows into Lake Powell at mouth

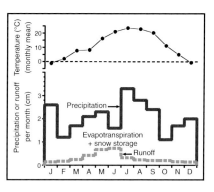

San Juan River, Utah (photo by Tim Palmer).

Virgin River

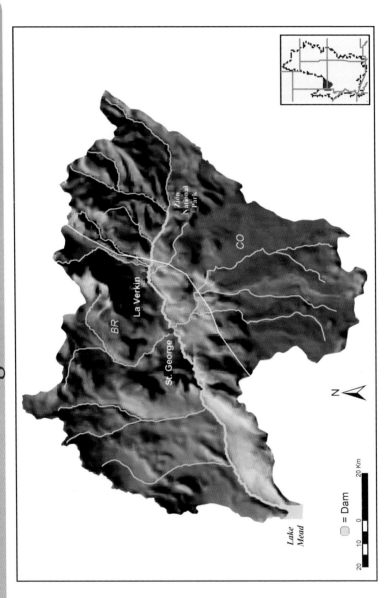

Relief: ~2635 m
Basin area: 13,200 km²
Mean discharge: 6.8 m³/s
Mean annual precipitation: 27 cm

Mean air temperature: 17.1°C
Mean water temperature: 17°C
No. of fish species: 21 (11 native)
No. of endangered species: 26

Physiographic provinces: Basin and Range (BR), Colorado Plateaus (CO)
Major fishes: black bullhead, desert sucker, flannelmouth sucker, green sunfish, mosquitofish, red shiner, speckled dace, Virgin River spinedace, woundfin
Major other aquatic vertebrates: American white pelican, beaver, belted kingfisher, Clark's grebe, great egret, lowland leopard frog, muskrat, northern leopard frog, osprey, snowy egret
Major benthic insects: mayflies (*Baetis*, *Tricorythodes*), caddisflies (*Cheumatopsyche*, *Chimarra*, *Hydropsyche*, *Nectopsyche*, *Ochrotrichia*), true flies (chironomid midges, black flies)
Nonnative species: black bullhead, bullfrog, green sunfish, red shiner, Russian olive, saltcedar, virile crayfish
Major riparian plants: box elder, coyote willow, Emory baccharis,

Fremont cottonwood, Russian olive, saltcedar
Special features: runs through Zion National Park to Lake Mead; highest water quality through Zion; recommended for conservation by World Wildlife Fund
Fragmentation: no dams but two major main-stem diversions include Quail Lake diversion near Virgin, Utah, and the Washington Fields diversion near Saint George, Utah

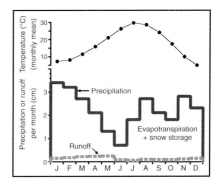

Virgin River above Virgin, Utah (photo by Tim Palmer).

Bill Williams River

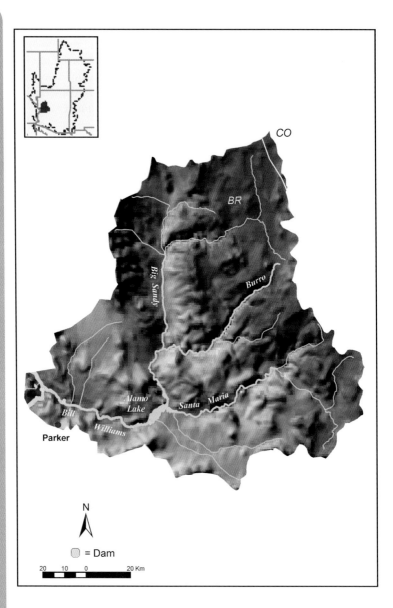

N

△

◯ = Dam

20 10 0   20 Km

Relief: ~1250 m   Mean air temperature: 22.4°C
Basin area: 13,950 km²   Mean water temperature: 20°C
Mean discharge: 4.3 m³/s   No. of fish species. 13 (1 native)
Mean annual precipitation: 13 cm   No. of endangered species: 1

Physiographic provinces: Basin and Range (BR), Colorado Plateaus (CO)
Major fishes: bluegill, common carp, fathead minnow, green sunfish, largemouth bass, mosquitofish, razorback sucker, red shiner, yellow bullhead
Major other aquatic vertebrates: Arizona toad, beaver, lowland leopard frog, muskrat, river otter, red-spotted toad, Sonoran Desert toad, spiny-spotted turtle
Major benthic insects: mayflies (*Baetis*), caddisflies (*Cheumatopsyche, Culoptila, Helicopsyche, Hydroptila, Leucotrichia, Nectopsyche, Ochrotrichia, Protoptila, Smicridea*)
Nonnative species: bluegill, common carp, fathead minnow, golden shiner, goldfish, green sunfish, largemouth bass, mosquitofish, redear sunfish, red shiner, red swamp crayfish, saltcedar, spiny-spotted turtle, yellow bullhead, *Daphnia lumholtzi* in Alamo Lake

Major riparian plants: broadleaf cattail, bulrush, coyote willow, Fremont willow, Goodding willow, narrowleaf cattail, saltcedar
Special features: runs through Bill Williams Wildlife Refuge; high bird density (≥335 species); dramatic lateral and vertical variations in microclimate
Fragmentation: almost entirely regulated by Alamo Dam; low to zero flows from July through October

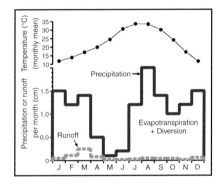

Bill Williams River, Arizona (photo by P. Shafroth).

Black River

Relief: ~1400 m
Basin area: 3400 km²
Mean discharge: 12 m³/s
Mean annual precipitation: 54 cm

Mean air temperature: 14°C
Mean water temperature: 15°C
No. of fish species: 13 (5 native)
No. of endangered species: 5

Physiographic provinces: Basin and Range (BR), Colorado Plateaus (CO)
Major fishes: Apache trout, brown trout, channel catfish, desert sucker, fathead minnow, rainbow trout, roundtail chub, smallmouth bass, Sonora sucker, speckled dace
Major other aquatic vertebrates: beaver, belted kingfisher, great blue heron, mallard, muskrat, osprey, river otter
Major benthic insects: mayflies (*Baetis*, *Epeorus Heptagenia*, *Tricorythodes*, *Thraulodes*), caddisflies (*Agapetus*, *Atopsyche*, *Brachycentrus*, *Ceratopsyche*, *Cheumatopsyche*, *Chimarra*, *Glossosoma*, *Helicopsyche*, *Hydropsyche*, *Lepidostoma*, *Limnephilus*, *Marilia*, *Micrasema*, *Phylloicus*), true flies (chironomid midges, black flies)
Nonnative species: bluegill, brown trout, bullfrog, channel catfish, cutthroat trout, fathead minnow, green sunfish, rainbow trout, smallmouth bass, saltcedar, virile crayfish
Major riparian plants: Bebb willow, coyote willow, Fremont cottonwood, Geyer willow, Goodding willow, narrowleaf cottonwood, thin-leaf alder, saltcedar
Special features: some of most natural riparian communities remaining in Arizona; passes through scenic Bear Wallow Wilderness Area and

Fort Apache and San Carlos Indian reservations
Fragmentation: relatively unfragmented; 3 small reservoirs in headwaters fed by tributaries during snowmelt

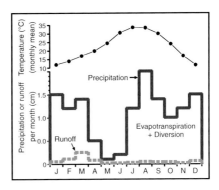

Black River at Wildcat Bridge, Arizona (photo by D. Blinn).

Verde River

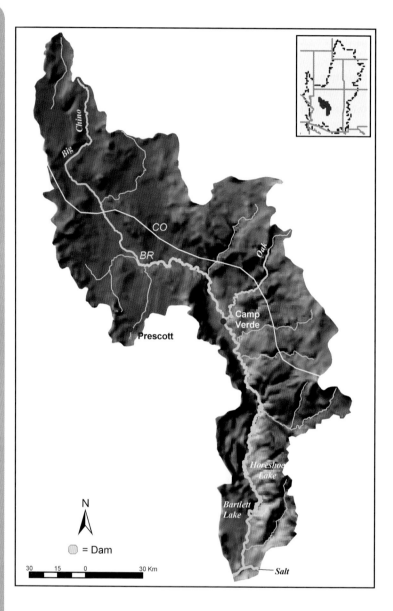

= Dam

N

30   15   0   30 Km

Relief: ~925 m
Basin area: 16,190 km^2
Mean discharge: 17 m^3/s
Mean annual precipitation: 35 cm

Mean air temperature: 20°C
Mean water temperature: 16.5°C
No. of fish species: 27 (10 native)
No. of endangered species: 8

Physiographic provinces: Basin and Range (BR), Colorado Plateaus (CO)
Major fishes: common carp, desert sucker, flathead catfish, Gila chub, green sunfish, headwater chub, largemouth bass, mosquitofish, red shiner, roundtail chub, small-mouth bass, Sonora sucker, yellow bullhead
Major other aquatic vertebrates: Arizona toad, beaver, belted kingfisher, bullfrog, Chiricahua leopard frog, great blue heron, lowland leopard frog, river otter, Sonoran mud turtle, striped chorus frog, Woodhouse toad
Major benthic insects: mayflies (*Baetis, Heptagenia*), caddisflies (*Cheumatopsyche, Chimarra, Helicopsyche, Heterelmis, Hydropsyche, Hydroptila, Ochrotrichia, Oecetis*)
Nonnative species: Asiatic clam, bullfrog, common carp, flathead catfish, green sunfish, largemouth bass, mosquitofish, red shiner, smallmouth bass, saltcedar, virile crayfish, yellow bullhead

Major riparian plants: Arizona alder, Arizona sycamore, Arizona walnut, arroyo willow, box elder, cattails, common reed, coyote willow, Fremont willow, Goodding willow, saltcedar
Special features: one of the largest perennial rivers in Gila basin; upper section recommended for conservation by World Wildlife Fund; 65 km National Wild and Scenic River
Fragmentation: lower third regulated by dams; 7 water-diversion dams

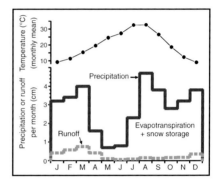

Verde River below Camp Verde, Arizona (photo by Tim Palmer).

Salt River

Relief: ~2540 m
Basin area: 35,480 km^2
Mean discharge: <25 m^3/s
Mean annual precipitation: 40 cm

Mean air temperature: 22°C
Mean water temperature: 19°C
No. of fish species: 16 (9 native)
No. of endangered species: 7

Physiographic provinces: Basin and Range (BR), Colorado Plateaus (CO)
Major fishes: bluegill, channel catfish, common carp, desert sucker, fathead minnow, flathead catfish, golden shiner, green sunfish, longfin dace, red shiner, Sonora sucker, speckled dace
Major other aquatic vertebrates: Arizona toad, bullfrog, Chiricahua leopard frog, northern leopard frog, lowland leopard frog, Sonoran mud turtle, virile crayfish
Major benthic insects: mayflies (*Baetis*, *Callibaetis*, *Tricorythodes*), caddisflies (*Hydropsyche*, *Hydroptila*), true flies (chironomid midges)
Nonnative species: bluegill, bullfrog, channel catfish, common carp, fathead minnow, flathead catfish, golden shiner, green sunfish, red shiner, Rio Grande leopard frog, saltcedar, virile crayfish

Major riparian plants: mesquite, saltcedar
Special features: lower section flows through scenic Salt River Canyon
Fragmentation: lower sections highly regulated by Roosevelt, Stuart Mountain (Saguaro Res.), Mormon Flat, and Horse Mesa dams; heavy use of diversion canals

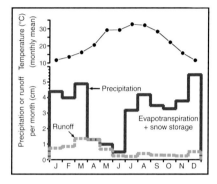

Salt River near Phoenix, Arizona (photo by C. E. Cushing).

Chapter 12

Pacific Coast Rivers of the Coterminous United States

James L. Carter and Vincent H. Resh

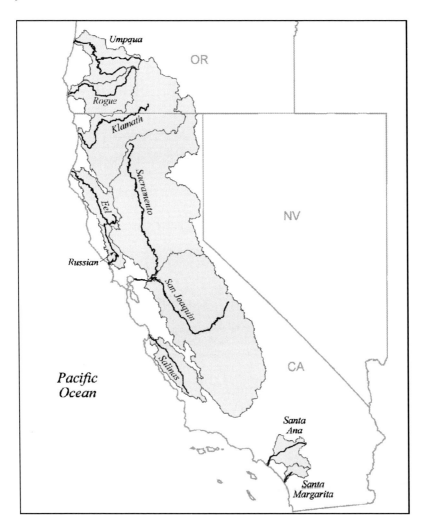

DOI: 10.1016/B978-0-12-375088-4.00012-9

The rivers discussed in this chapter are mainly located in the southern portion of the Pacific Mountain System and discharge into the Pacific Ocean (see map). Located from south of the Columbia River to southern California, these river basins occupy an area just over 10° of latitude and 10° of longitude. The rivers and the biota have been influenced more than any other rivers in North America by tectonic activity that is both geologically recent and ongoing, including: 1) periods of mountain building through uplift, volcanism, and accretion; 2) periods of massive erosion; 3) changes in sea level; and 4) large-scale faulting. These factors have influenced the aspect and gradient of the rivers, and created a topography that affects sub-regional precipitation patterns, which in turn influences the hydrology, geomorphology, and present-day biology of the region's rivers. The importance of the recency of these events is shown by the uniqueness and high endemism of the regional flora and fauna.

From the time of the arrival of the first people to these Pacific basins, often estimated to have occurred 11,000 years ago, rivers have provided these humans with food, water, and transportation. The diversity of freshwater and estuarine species, as well as the abundant terrestrial wildlife and plants, supported some of the highest densities of Native Americans in North America. Spanish exploration began in the early to mid-16th century and by the 1700s, numerous settlements were established along the Pacific coast. The defining event in the human history of this region was John Marshall's discovery of gold at Sutter's Mill, California, in 1848. Immigration associated with this event significantly increased the population (100,000 miners had arrived by 1850). The negative effect was the excess sediment from hydraulic mining that flowed into stream channels and the contamination from mercury that was used to extract gold; these represented the first great assault on the region's rivers. However, today the most significant human influences on rivers are from impoundments and withdrawals, with the water principally (~80%) used for irrigation.

Numerous major rivers are found in the southern portion of the Pacific Mountain system. The largest river of the region (by discharge) is the Sacramento with a basin area of more than $72,000\,km^2$ and a mean discharge of $657\,m^3$. Joining it from the south is the San Joaquin, which drains an even larger basin ($>83,000\,km^2$), but has a substantially lower mean discharge of $132\,m^3/s$. Other rivers in Northern California (Eel, Klamath) and Oregon (Rogue, Umpqua) have mean discharges $>200\,m^3/s$. Features of the Salinas, Russian, Santa Ana, and Santa Margarita rivers are also described.

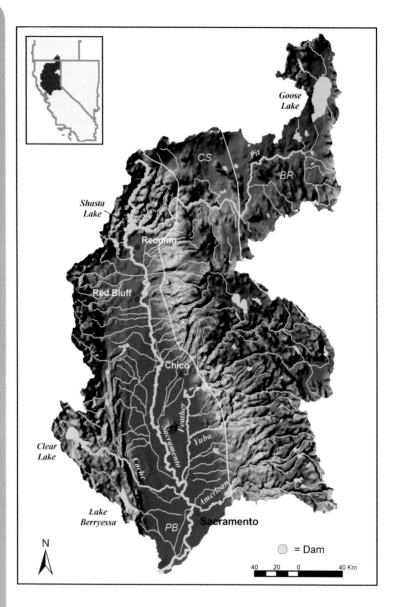

Sacramento River

Relief: 4317 m   Mean air temperature: 12.9°C
Basin area: 72,132 km²   Mean water temperature: NA
Mean discharge: 657 m³/s   No. of fish species: 69 (29 native)
Mean annual precipitation: 90 cm No. of endangered species: >50

Physiographic provinces: Pacific Border (PB), Cascades–Sierra Nevada Mountains (CS), Basin and Range (BR)
Major fishes: Pacific lamprey, river lamprey, white sturgeon, green sturgeon, Sacramento blackfish, hardhead, hitch, Sacramento pikeminnow, tui chub, Sacramento splittail, California roach, speckled dace, Sacramento sucker, delta smelt, longfin smelt, Modoc sucker, chinook salmon, rainbow trout, coho salmon, threespine stickleback, Sacramento perch, tule perch, staghorn sculpin, riffle sculpin
Major other aquatic vertebrates: California newt, Sierra newt, California red-legged frog, foothill yellow-legged frog, mountain yellow-legged frog, western leopard frog, western toad, Yosemite toad, western pond turtle, western aquatic garter snake, water shrew, mountain beaver, beaver, muskrat, raccoon, river otter, ermine (short-tailed weasel), long-tailed weasel, mink
Major benthic insects: mayflies (*Siphlonurus*, *Baetis*, *Drunella*, *Ephemerella*, *Serratella*, *Epeorus*, *Rhithrogena*, *Cinygmula*), stoneflies (*Pteronarcys*, *Malenka*, *Zapada*, *Hesperoperla*, *Calineuria*, *Isoperla*, *Suwallia*, *Sweltsa*, *Eucapnopsis*), caddisflies (*Arctopsyche*, *Hydropsyche*, *Cheumatopsyche*, *Rhyacophila*, *Hydroptila*)
Nonnative species: eastern oyster, bullfrog, >40 species of fishes (e.g., American shad, striped bass, and many warmwater fishes)
Major riparian plants: arroyo willow, black willow, narrowleaf willow, Pacific willow, red willow, black cottonwood, Fremont cottonwood, California sycamore, mulefat,

mountain alder, white alder, buttonbush, water birch
Special features: largest river in California; several National Wild and Scenic reaches; one of largest salmonid runs in California; discharges through the highly productive Sacramento–San Joaquin Delta into San Francisco Bay
Fragmentation: hundreds of dams and water withdrawals within basin

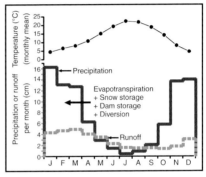

Sacramento River below Chico, California (photo by Tim Palmer).

San Joaquin River

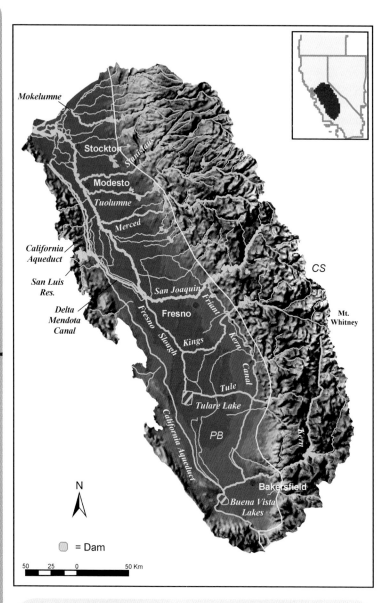

Relief: 4418 m
Basin area: 83,409 km^2
Mean discharge: 132 m^3/s
Mean annual precipitation: 49 cm

Mean air temperature: 15.7°C
Mean water temperature: NA
No. of fish species: 63
(native and nonnative)
No. of endangered species: >50

Physiographic provinces: Pacific Border (PB), Cascade–Sierra Nevada Mountains (CS)

Major fishes: Pacific lamprey, white sturgeon, Sacramento blackfish, hardhead, hitch, Sacramento pikeminnow, California roach, speckled dace, Sacramento sucker, delta smelt, longfin smelt, chinook salmon, rainbow trout, threespine stickleback, Sacramento perch, tule perch, staghorn sculpin, riffle sculpin

Major other aquatic vertebrates: California newt, Sierra newt, tailed frog, California red-legged frog, foothill yellow-legged frog, mountain yellow-legged frog, bullfrog, western leopard frog, western pond turtle, western aquatic garter snake, water shrew, mountain beaver, beaver, muskrat, nutria, raccoon, river otter, ermine, long-tailed weasel, mink

Major benthic insects: mayflies (*Baetis, Epeorus, Caenis, Heptagenia, Fallceon, Ephemerella, Drunella*), stoneflies (*Hesperoperla, Suwallia, Eucapnopsis*), caddisflies (*Hydropsyche, Arctopsyche, Rhyacophila, Nectopsyche*)

Nonnative species: bullfrog, >40 species of fishes (largemouth bass, smallmouth bass, crappie, bluegill)

Major riparian plants: arroyo willow, black willow, narrowleaf willow, Pacific willow, red willow, black cottonwood, Fremont cottonwood, buttonbush, California sycamore, mulefat, white alder

Special features: begins in high Sierra Nevada; several tributaries designated National Wild and Scenic Rivers; flows through intensive agricultural region before discharging through highly productive Sacramento–San Joaquin Delta into San Francisco Bay

Fragmentation: hundreds of dams and water withdrawals within basin

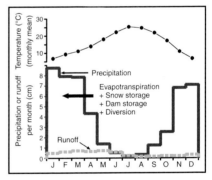

San Joaquin River, within the Central Valley, California (photo courtesy of Great Valley Museum, Modesto, California).

Salinas River

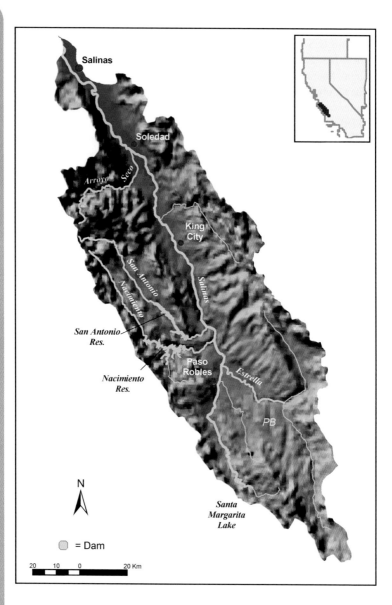

Salinas

Soledad

Arroyo Seco

King City

San Antonio

Nacimiento

Salinas

San Antonio Res.

Nacimiento Res.

Paso Robles

Estrella

PB

N

Santa Margarita Lake

○ = Dam

20 10 0   20 Km

Relief: 1787 m
Basin area: 10,983 km^2
Mean discharge: 12.7 m^3/s
Mean annual precipitation: 36 cm

Mean air temperature: 14.6°C
Mean water temperature: NA
No. of fish species: 36 (16 native)
No. of endangered species: 42

Physiographic province: Pacific Border (PB)

Major fishes: Pacific lamprey, Pacific brook lamprey, Sacramento blackfish, hitch, Sacramento pikeminnow, California roach, speckled dace, Sacramento sucker, coho salmon, steelhead, rainbow trout, threespine stickleback, tidewater goby, staghorn sculpin, coastrange sculpin, prickly sculpin, riffle sculpin

Major other aquatic vertebrates: California giant salamander, California newt, Pacific chorus frog, California chorus frog, California red-legged frog, foothill yellow-legged frog, western toad, arroyo toad, western pond turtle, aquatic/Santa Cruz garter snake, beaver, muskrat, raccoon, long-tailed weasel

Major benthic insects: mayflies (*Acentrella, Baetis, Diphetor, Fallceon, Centroptilum, Procloeon, Serratella, Tricorythodes*), caddisflies (*Hydropsyche, Cheumatopsyche*)

Nonnative species: bullfrog, threadfin shad, common carp, golden shiner, fathead minnow, channel catfish, white catfish, brown bullhead, black bullhead, brown trout, mosquitofish, inland silverside, white bass, black crappie, white crappie, green sunfish, bluegill, redear sunfish, largemouth bass, smallmouth bass

Major riparian plants: arroyo willow, narrowleaf willow, red willow, Sitka willow, buttonbush, California sycamore, Fremont cottonwood, mulefat, white alder

Special features: flows through the Salinas Valley, which is often referred to as "America's Salad Bowl"; under natural flow conditions it is one of the longest underground rivers in North America

Fragmentation: 17 dams

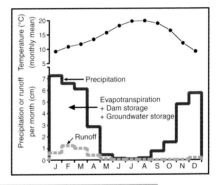

Salinas River, California, near U.S. highway 101 (on right) (photo by K. Ekelund, Monterey County Water Resources Agency).

Klamath River

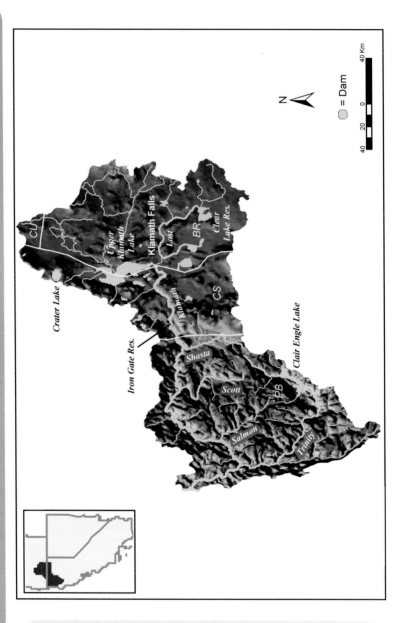

Relief: 2894 m
Basin area: 40,608 km²
Mean discharge: 501 m³/s
Mean annual precipitation: 85 cm

Mean air temperature: 10.5°C
Mean water temperature: NA
No. of fish species: 48 (30 native)
No. of endangered species: 41

Physiographic provinces: Pacific Border (PB), Basin and Range (BR), Cascade–Sierra Nevada Mountains (CS)

Major fishes: Pacific lamprey, Klamath River lamprey, white sturgeon, green sturgeon, blue chub, tui chub, speckled dace, Lost River sucker, Klamath smallscale sucker, longfin smelt, chum salmon, coho salmon, chinook salmon, rainbow trout, cutthroat trout, steelhead, several sculpins

Major other aquatic vertebrates: northwestern salamander, southern torrent salamander, rough skinned newt, red-legged frog, California red-legged frog, foothill yellow-legged frog, Cascades frog, spotted frog, beaver, muskrat, nutria, river otter, mink

Major benthic insects: mayflies (*Baetis, Drunella, Rhithrogena, Cinygmula, Epeorus, Ironodes, Paraleptophlebia*), stoneflies (*Yoraperla, Calineuria, Hesperoperla, Zapada, Malenka*), caddisflies (*Hydropsyche, Nectopsyche, Oecetis, Helicopsyche, Glossosoma, Lepidostoma, Agapetus*)

Nonnative species: bullfrog, American shad, goldfish, golden shiner, fathead minnow, brown bullhead, black bullhead, wakasagi, kokanee, brown trout, brook trout, brook stickleback, Sacramento perch, crappie

Major riparian plants: arroyo willow, Hooker willow, black willow, narrowleaf willow, Pacific willow, red willow, sandbar willow, Sitka willow, black cottonwood, Fremont cottonwood

Special features: flows through rugged, species-rich Klamath Mountains; below most downstream dam flows unimpounded for >450 km to Pacific; >900 km designated as National Wild and Scenic River

Fragmentation: 24 dams

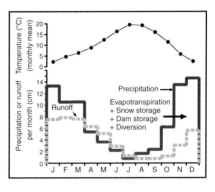

Klamath River between the towns of Seiad Valley and Happy Camp, California (photo by Steve Fend, USGS).

Rogue River

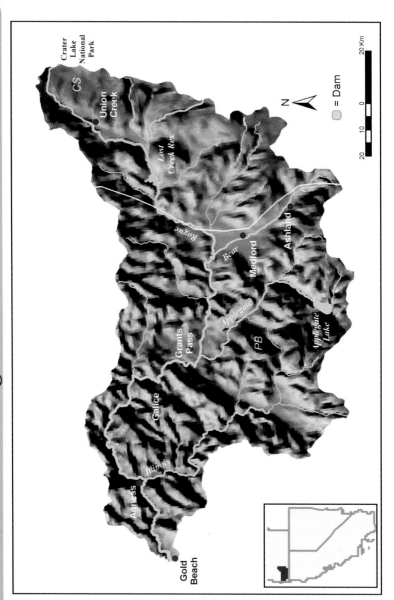

Relief: 2894 m
Basin area: 13,348 km²
Mean discharge: 285 m³/s
Mean annual precipitation: 97 cm

Mean air temperature: 11.6°C
Mean water temperature: NA
No. of fish species: 23 (14 native)
No. of endangered species: 11

Physiographic provinces: Cascade–Sierra Nevada Mountains (CS), Pacific Border (PB)

Major fishes: Pacific lamprey, green sturgeon, white sturgeon, coastal cutthroat, pink salmon, coho salmon, rainbow trout, Chinook salmon, speckled dace, tui chub, Klamath smallscale sucker, steelhead

Major other aquatic vertebrates: northwestern salamander, Pacific giant salamander, southern torrent salamander, rough skinned newt, red-legged frog, foothill yellow-legged frog, Cascades frog, spotted frog, western toad, western pond turtle, western aquatic garter snake, water shrew, mountain beaver, beaver, muskrat, raccoon, river otter, ermine, long-tailed weasel, mink

Major benthic insects: mayflies (*Baetis, Diphetor, Acentrella, Ephemerella, Drunella, Rhithrogena, Epeorus, Paraleptophlebia*), stoneflies (*Zapada, Calineuria, Malenka, Sweltsa*), caddisflies (*Hydropsyche, Cheumatopsyche, Arctopsyche, Parapsyche, Lepidostoma, Rhyacophila, Micrasema, Glossosoma*)

Nonnative species: brook trout, common carp, golden shiner, Umpqua pikeminnow, redside shiner, brown bullhead, smallmouth bass, largemouth bass, black crappie, bullfrog

Major riparian plants: sandbar willow, Geyer willow, Pacific willow, yellow willow, Scouler willow, white alder, red alder, black cottonwood, Oregon ash

Special features: begins in high Cascade Mountains on slopes of Crater Lake National Park; one of first National Wild and Scenic Rivers; one of best salmonid fisheries in west; renowned for white-water boating

Fragmentation: ~80 dams

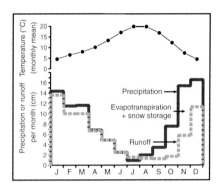

Rogue River, Oregon (photo by Tim Palmer).

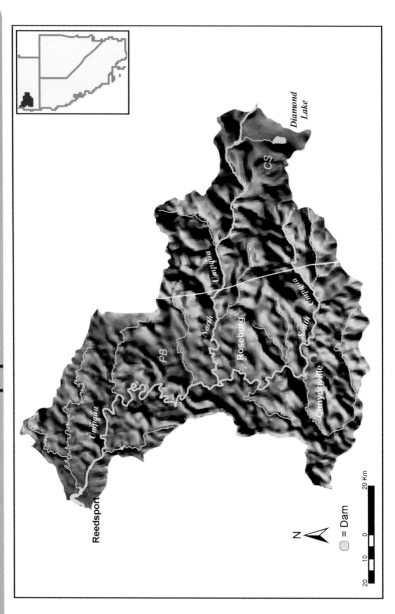

Umpqua River

Relief: 2799 m
Basin area: 12,133 km^2
Mean discharge: 211 m^3/s
Mean annual precipitation: 115 cm

Mean air temperature: 11.7°C
Mean water temperature: NA
No. of fish species: 27 (19 native)
No. of endangered species: 9

Physiographic provinces: Cascade–Sierra Nevada Mountains (CS), Pacific Border (PB)
Major fishes: river lamprey, western brook lamprey, Pacific lamprey, coastal cutthroat, chum salmon, coho salmon, rainbow trout, chinook salmon, Umpqua chub, Umpqua pikeminnow, Umpqua dace, speckled dace, tui chub, largescale sucker, threespine stickleback, coastrange sculpin, prickly sculpin, riffle sculpin, reticulate sculpin
Major other aquatic vertebrates: northwestern salamander, Pacific giant salamander, southern torrent salamander, rough skinned newt, Pacific chorus frog, red-legged frog, foothill yellow-legged frog, Cascades frog, spotted frog, western toad, western aquatic garter snake, water shrew, mountain beaver, beaver, muskrat, nutria, raccoon, river otter, mink
Major benthic insects: mayflies (*Baetis tricaudatus*, *Diphetor*, *Paraleptophlebia*, *Rhithrogena*), stoneflies (*Calineuria*, *Hesperoperla*, *Sweltsa*, *Malenka*, *Zapada*), caddisflies (*Hydropsyche*, *Glossosoma*, *Neophylax*)

Nonnative species: American shad, redside shiner, yellow bullhead, western mosquitofish, bluegill, smallmouth bass, largemouth bass, black crappie, brook trout, bullfrog
Major riparian plants: sandbar willow, Geyer willow, Pacific willow, yellow willow, Scouler willow, white alder, red alder, black cottonwood, Oregon ash
Special features: begins in Cascades; world renowned steelhead and salmon fishery; portions have National Wild and Scenic River status
Fragmentation: 64 dams in Douglas County, Oregon; fewest dams of any large basin in Oregon

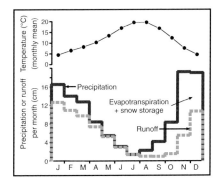

Umpqua River at Elkton, Oregon (photo by Tim Palmer).

Eel River

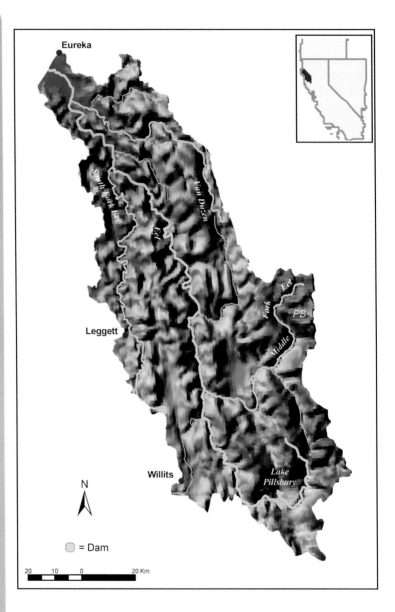

Eureka

South Fork Eel

Van Duzen

Eel

Eel

Fork

PB

Middle

Leggett

Willits

Lake Pillsbury

N

◯ = Dam

20   10   0   20 Km

Relief: 2270 m   Mean air temperature: 12.7°C
Basin area: 9456 km²   Mean water temperature: NA
Mean discharge: 210 m³/s   No. of fish species: 25 (15 native)
Mean annual precipitation: 133 cm   No. of endangered species: 12

Physiographic province: Pacific Border (PB)

Major fishes: Pacific lamprey, river lamprey, Pacific brook lamprey, Sacramento sucker, coho salmon, chinook salmon, rainbow trout, cutthroat trout, threespine stickleback, staghorn sculpin, coastrange sculpin, prickly sculpin

Major other aquatic vertebrates: northwestern salamander, Pacific giant salamander, southern torrent salamander, rough skinned newt, California newt, red-bellied newt, red-legged frog, foothill yellow-legged frog, western toad, western pond turtle, western aquatic garter snake, aquatic/Santa Cruz garter snake, beaver, muskrat, raccoon, river otter, mink

Major benthic insects: mayflies (*Baetis, Ephemerella, Epeorus, Cinygmula, Rhithrogena, Paraleptophlebia*), stoneflies (*Malenka, Calineuria, Sweltsa*), caddisflies (*Hydropsyche, Rhyacophila, Helicopsyche*)

Nonnative species: bullfrog, American shad, threadfin shad, golden shiner, Sacramento pikeminnow, California roach, speckled dace, fathead minnow, brown bullhead, green sunfish, bluegill

Major riparian plants: arroyo willow, Hooker willow, black willow, narrowleaf willow, Pacific willow, red willow, sandbar willow, Sitka willow, black cottonwood, Fremont cottonwood, mulefat, white alder

Special features: begins in Six Rivers and Mendocino National Forests; flows past some of the most spectacular ancient redwood groves in California; more kilometers designated as National Wild and Scenic Rivers than any basin in California

Fragmentation: 14 dams

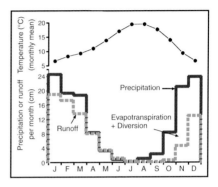

South Fork of Eel River in Humboldt Redwoods, California (photo by Tim Palmer).

Russian River

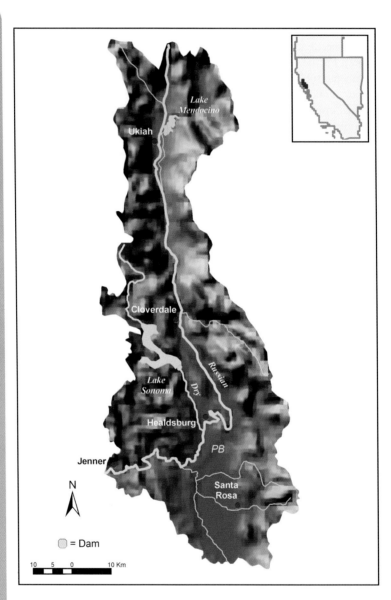

Relief: 1324 m   Mean air temperature: 14.9°C
Basin area: 3728 km²   Mean water temperature: NA
Mean discharge: 66 m³/s   No. of fish species: 41 (20 native)
Mean annual precipitation: 105 cm   No. of endangered species: 43

Physiographic province: Pacific Border (PB)
Major fishes: Pacific lamprey, river lamprey, Pacific brook lamprey, hardhead, hitch, Sacramento pikeminnow, California roach, Sacramento sucker, Klamath largescale sucker, longfin smelt, coho salmon, chinook salmon, rainbow trout, threespine stickleback, tule perch, staghorn sculpin, coastrange sculpin, prickly sculpin, riffle sculpin
Major other aquatic vertebrates: northwestern salamander, Pacific giant salamander, California giant salamander, southern torrent salamander, rough skinned newt, California newt, sierra newt, red-bellied newt, foothill yellow-legged frog, mountain yellow-legged frog, western aquatic garter snake, beaver, muskrat, mink
Major benthic insects: mayflies (*Baetis, Diphetor, Tricorythodes, Drunella, Ephemerella, Paraleptophlebia, Rhithrogena, Epeorus*), stoneflies (*Calineuria, Malenka, Sweltsa*), caddisflies (*Hydropsyche, Rhyacophila, Lepidostoma*)

Nonnative species: bullfrog, American shad, threadfin shad, common carp, goldfish, golden shiner, Sacramento blackfish, fathead minnow, green sunfish, bluegill, several catfish, brown bullhead, black bullhead, brown trout, mosquitofish, inland silverside, crappie, redear sunfish, largemouth bass, smallmouth bass
Major riparian plants: arroyo willow, Hooker willow, narrowleaf willow, Pacific willow, red willow, sandbar willow, Sitka willow
Special features: classic Mediterranean-type river in California
Fragmentation: 62 dams

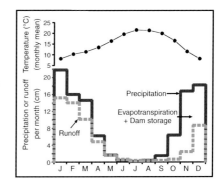

Russian River, California (photo by Tim Palmer).

Santa Ana River

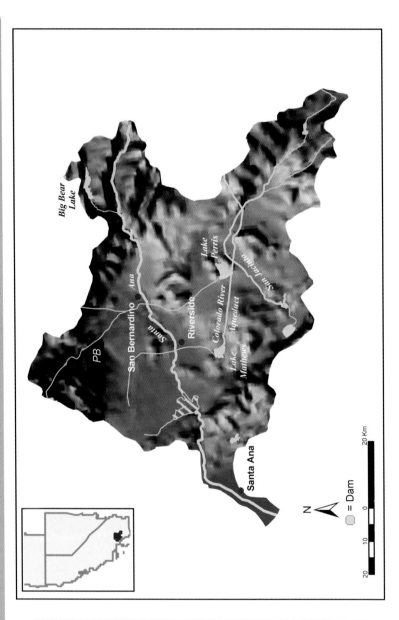

Relief: 3506 m

Basin area: 6314 km²

Mean discharge: 1.7 m³/s

Mean annual precipitation: 34 cm

Mean air temperature: 17.2°C

Mean water temperature: NA

No. of fish species: 45 (9 native)

No. of endangered species: 54

Physiographic province: Pacific Border (PB)

Major fishes: arroyo chub, speckled dace, rainbow trout, California killifish, threespine stickleback, striped mullet, tidewater goby, staghorn sculpin, prickly sculpin

Major other aquatic vertebrates: California newt, Sierra newt, California red-legged frog, mountain yellow-legged frog, western leopard frog, African clawed frog, western toad, arroyo toad, western pond turtle, western aquatic garter snake, beaver, muskrat, raccoon, river otter, long-tailed weasel

Major benthic insects: mayflies (*Baetis, Fallceon, Caudatella, Drunella, Tricorythodes*), caddisflies (*Hydropsyche, Helicopsyche*)

Nonnative species: bullfrog, threadfin shad, common carp, goldfish, hitch, Sacramento pikeminnow, Colorado pikeminnow, red shiner, fathead minnow, catfish, brown trout, rainwater killifish, mosquitofish, sailfin molly, inland silverside, crappie, green sunfish, bluegill, pumpkinseed,

redear sunfish, largemouth bass, spotted bass, smallmouth bass, bigscale logperch, redbelly tilapia, tule perch, yellowfin goby, Shimofuri goby; list includes Los Angeles basin

Major riparian plants: arroyo willow, black willow, narrowleaf willow, Pacific willow, red willow, California sycamore, Fremont cottonwood, mulefat, white alder

Special features: one of the largest river systems in southern California, but intermittent, channelized, and highly urbanized; upper portion still retains natural characteristics

Fragmentation: 52 dams

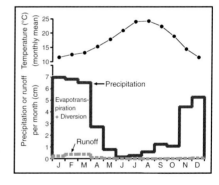

Santa Ana River downstream of Highway 90, California (photo by C. Burton, USGS).

Santa Margarita River

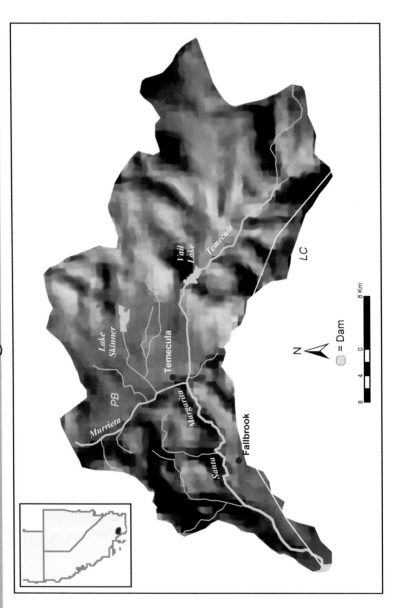

Relief: 2076 m

Basin area: 1896 km^2

Mean discharge: 1.2 m^3/s

Mean annual precipitation: 50 cm

Mean air temperature: 14.6°C

Mean water temperature: N.A.

No. of fish species: 17 (6 native)

No. of endangered species: 52

Physiographic provinces: Lower California (LC), Pacific Border (PB)
Major fishes: arroyo chub, rainbow trout, California killifish, striped mullet, longjaw mudsucker, staghorn sculpin
Major other aquatic vertebrates: California newt, Pacific chorus frog, California chorus frog, California red-legged frog, mountain yellow-legged frog, western leopard frog, African clawed frog, western toad, arroyo toad, western pond turtle, western aquatic garter snake, beaver, muskrat, raccoon, long-tailed weasel
Major benthic insects: mayflies (*Baetis, Tricorythodes, Fallceon, Centroptilum/Procloeon*), stoneflies (*Isoperla, Malenka, Zapada*), caddisflies (*Hydropsyche, Cheumatopsyche, Amiocentrus, Micrasema*), true flies (*Simulium*)
Nonnative species: bullfrog, common carp, channel catfish, black bullhead, yellow bullhead, mosquitofish, black crappie, white crappie, green sunfish, bluegill, largemouth bass, redeye bass, yellowfin goby
Major riparian plants: arroyo, black, narrowleaf, Pacific, and red willow; California sycamore; Fremont cottonwood; mulefat; white alder
Special features: one of the last free-flowing rivers in southern California; one of the most intact and continuous riparian corridors in region
Fragmentation: 9 dams

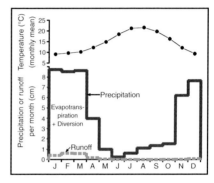

Lower portion of Santa Margarita River, California (photo by Cheryl S. Brehme, USGS).

Chapter 13

Columbia River Basin

Jack A. Stanford, F. Richard Hauer, Stanley V. Gregory, and Eric B. Snyder

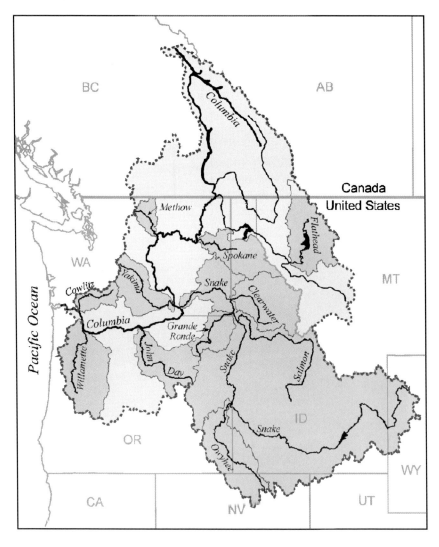

The Columbia River drains 724,025 km² of the American Pacific Northwest, including parts of Washington, Oregon, Nevada, Utah, Idaho, Wyoming, Montana, and British Columbia (see map). Landscape diversity is high, with the river draining parts of seven physiographic provinces, such as the Rocky Mountains in Canada and the Colombia-Snake River Plateaus. The Columbia obtains much of its runoff from distant Rocky Mountain headwaters, flows through a huge and mostly dry interior basin, gains significant runoff from the eastern slopes of the Cascade Mountains, squeezes through a narrow gap in the Cascade cordillera, picks up the Willamette River and softly passes through cool rainforests of the low Pacific coast mountains. With its enormous discharge of 7730 m³/s, the Columbia is a great water road for salmon and people; a purveyor of energy and irrigation water that has figured decisively in the development of the nation.[1]

Hunter-gatherer people have an ancient tenure in the Columbia River Basin, with the earliest evidence of human occupation corresponding to the last catastrophic floods from glacial outwash approximately 12,000 years ago.[2] Archaeological data indicate that from about 9000 years ago the resources of the Columbia River and its tributaries were important to early cultural systems. From about 4000 years ago through the early historic period, salmon in association with gathered roots and hunting formed the basis of the human economy in the Columbia Basin. When Boston fur trader and whaler Robert Gray first viewed and recorded the mouth of the Columbia in 1792, he had no idea that the river system was populated by over 100,000 Native Americans including Nez Perce, Salish, Kootenai, Umatilla, and Yakama people with strong Pacific coastal affinity, along with several tribes such as Blackfeet and Shoshone of Great Plains origin using the headwaters. The great expedition of Lewis and Clark (1804 to 1806) clearly revealed the physiographic breadth and ecological scope of the Columbia. Within 150 years, however, the rapids were tamed and the cultural and ecological landscape dramatically altered.

This chapter describes the Columbia main stem, several large tributaries, as well as smaller tributaries that reflect the diversity of the basin. The largest is the Snake/Salmon river system, beginning far inland in northwestern Wyoming, draining a basin area of 281,000 km², and having a mean discharge of 1565 m³/s. Several other rivers contribute >200 m³/s, including the Flathead, Willamette, Clearwater, Spokane, and Cowlitz. Other described rivers include the Owyhee, Methow, Yakima, Grande Ronde, and John Day.

---

1 White, R. 1995. The Organic Machine: The Remaking of the Columbia River. Hill and Wang, New York.

2 Attwater, B. S. 1986. Pleistocene glacial-lake deposits of the Sanpoil River Valley, Northeastern Washington. U.S. Geological Survey Bulletin 1661. U.S. Government Printing Office, Washington, D.C.

Columbia River

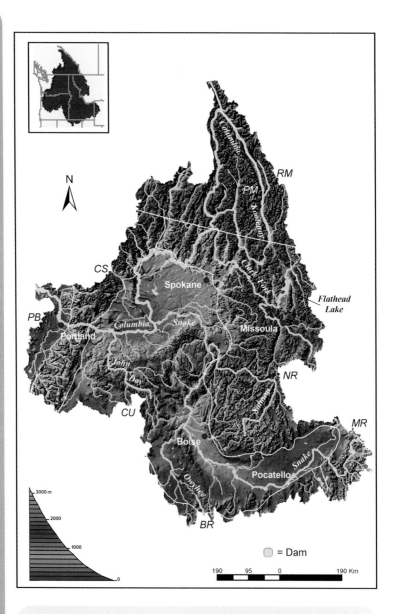

N

RM

PM

Kootenay

Columbia

CS

Spokane

Clark Fork

Flathead
Lake

PB

Columbia   Snake   Missoula

Portland

John

Day

NR

CU

Salmon

MR

Boise

Snake

Pocatello

Owyhee

3000 m

2000

1000

0

BR

◯ = Dam

190   95   0   190 Km

Relief: 4392 m
Basin area: 724,025 km²
Mean discharge: 7730 m³/s
Mean annual precipitation: 70 cm

Mean air temperature: 9.7°C
Mean water temperature: 13.3°C
No. of fish species: 103 (53
native) resident, 15 (12 native)
anadromous, 4 marine
No. of endangered species: 15

Physiographic provinces: Northern Rocky Mountains (NR), Middle Rocky Mountains (MR), Rocky Mountains in Canada (RM), Columbia–Snake River Plateaus (CU), Cascade–Sierra Mountains (CS), Pacific Border (PB), Basin and Range (BR)

Major fishes (main stem): chinook salmon, steelhead, northern pikeminnow, largescale sucker, mountain whitefish, eulachon, sculpins, speckled dace, chiselmouth, American shad

Major other aquatic vertebrates (main stem): beaver, painted turtle, bullfrog, muskrat, nutria, American merganser, northwest pond turtle, northern leopard frog, chorus frog, bald eagle, Canada goose, osprey, mink, river otter

Major benthic insects (main stem): caddisflies (*Glossosoma*, *Cheumatopsyche*, *Hydropsyche*, *Hydroptila*), stoneflies (*Arcynopteryx*), mayflies (*Baetis*, *Ephemerella*, *Ephemera*, *Ephoron*, *Heptagenia*, *Stenonema*, *Tricorythodes*)

Nonnative species (main stem): American shad, brown trout, lake trout, lake whitefish, grass carp, goldfish, tench, channel catfish, several bullheads, mosquitofish, largemouth bass, smallmouth bass, crappie, bluegill, pumpkinseed, walleye, yellow perch, northern pike

Major riparian plants (main stem): black cottonwood, Russian olive, western hemlock, water hemlock, box elder, alder, willow, red-osier dogwood, reed canary grass, cattail, bulrush, sedges, purple loosestrife

Special features: fourth-largest river flowing to ocean in North America; 62 subbasins in seven states and British Columbia

Fragmentation: main-stem river almost completely impounded by large dams

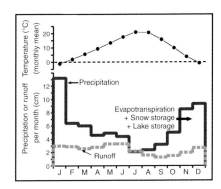

Columbia River Gorge (photo by A. C. Benke).

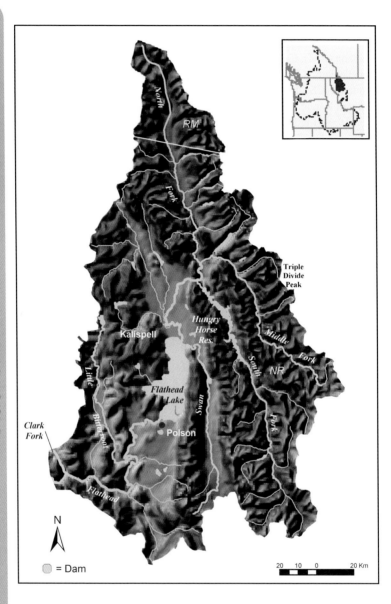

Flathead River

Relief: 1676 m
Basin area: 22,241 km²
Mean discharge: 340 m³/s
Mean annual precipitation: 56 cm

Mean air temperature: 8.6°C
Mean water temperature: 11.1°C
No. of fish species: 29 (12 native)
No. of endangered species: 2

Physiographic provinces: Northern Rocky Mountains (NR), Rocky Mountains in Canada (RM)
Major fishes: bull trout, westslope cutthroat trout, lake trout, mountain whitefish, lake whitefish, pigmy whitefish, sculpin, peamouth chub, longnose sucker, largescale sucker, northern pikeminnow
Major other aquatic vertebrates: spotted frog, boreal toad, painted turtle, tailed frog, three-toed salamander, beaver, river otter, mink, osprey, bald eagle
Major benthic insects: caddisflies (*Parapsyche*, *Arctopsyche*, *Hydropsyche*, *Cheumatopsyche*, *Glossosoma*, *Brachycentrus*, *Rhyacophila*), stoneflies (*Taeniopteryx*, *Pteronarcys*, *Pteronarcella*, *Hesperoperla*, *Claassenia*, *Isocapnia*, *Paraperla*), mayflies (*Rhithrogena*, *Baetis*, *Ephemerella*, *Drunella*)
Nonnative species: lake trout, lake whitefish, kokanee, yellow perch, northern pike, rainbow

trout, brook trout, largemouth bass, smallmouth bass, virile crayfish
Major riparian plants: black cottonwood, green alder, willow, red-osier dogwood, Englemann spruce, beaked sedge
Special features: National Wild and Scenic River segments; 42% of upper basin designated national park or wilderness; Flathead Indian reservation; 3 National Wildlife Refuges; Flathead Lake largest (surface area) lake in western United States
Fragmentation: 2 major dams

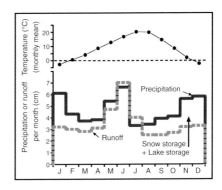

Flathead River near Bad Rock Canyon (photo by C. E. Cushing).

Snake/Salmon River

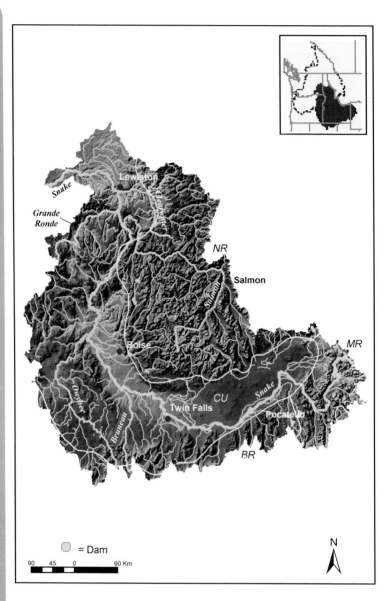

= Dam

90   45   0   90 Km

N

Relief: 3048 m

Relief: 3048 m
Basin area: 281,000 km²
Mean discharge: 1565 m³/s
Mean annual precipitation: 36 cm

Mean air temperature: 9.1°C
Mean water temperature: 11.5°C
No. of fish species: 39 (19 native)
No. of endangered species: 12

Physiographic provinces: Columbia–Snake River Plateaus (CU), Northern Rocky Mountains (NR), Middle Rocky Mountains (MR), Basin and Range (BR)

Major fishes: Yellowstone cutthroat trout, chinook salmon, steelhead, rainbow trout, bull trout, mountain whitefish, chiselmouth, northern pikeminnow, longnose dace, speckled dace, Utah chub, redside shiner, largescale sucker, sculpin

Major other aquatic vertebrates: northern leopard frog, Columbia spotted frog, tailed frog, western painted turtle, wood duck, Canada goose, trumpeter swan, sandhill crane, white pelican, bald eagle, osprey, beaver, river otter, muskrat

Major benthic insects: caddisflies (*Brachycentrus, Glossosoma, Arctopsyche, Cheumatopsyche, Hydropsyche*), mayflies (*Baetis, Drunella, Ephemerella, Rhithrogena, Tricorythodes*), stoneflies (*Sweltsa, Zapada, Claassenia, Hesperoperla, Skwala*), true flies (*Simulium, Tipula*)

Nonnative species: smallmouth bass, walleye, largemouth bass, carp, brown bullhead, black crappie, mosquitofish, pumpkinseed, yellow perch, brown trout, brook trout, channel catfish, black bullhead, bluegill, New Zealand mudsnail

Major riparian plants: black cottonwood, red-osier dogwood, willow, alder, purple loosestrife

Special features: headwaters of Snake in Yellowstone and Grand Teton National Parks; several tributaries are National Wild and Scenic Rivers

Fragmentation: several major dams

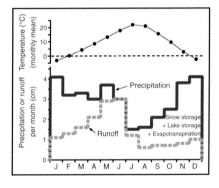

Snake River below confluence with Grande Ronde River, Washington (photo by C. E. Cushing).

Yakima River

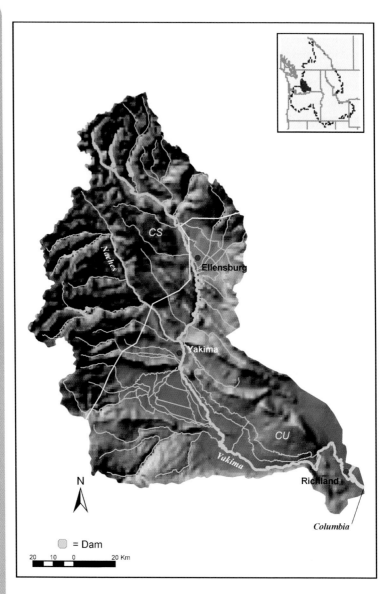

= Dam

20   10   0   20 Km

Relief: 2334 m   Mean air temperature: 12.3°C
Basin area: 15,900 km²   Mean water temperature: 13.3°C
Mean discharge: 102 m³/s   No. of fish species: 50
Mean annual precipitation: 19 cm   No. of endangered species: 2

Physiographic provinces: Cascade–
Sierra Mountains (CS), Columbia–
Snake River Plateaus (CU)
Major fishes: chinook salmon,
steelhead, rainbow trout, mountain
whitefish, chiselmouth, carp, northern
pikeminnow, longnose dace, speckled
dace, redside shiner, bridgelip sucker,
largescale sucker, sculpin
Major other aquatic vertebrates:
northern leopard frog, Columbia
spotted frog, northwest salamander,
Pacific giant salamander, western
pond turtle, wood duck, mallard duck,
Canada goose, tundra swan, trumpeter
swan, sandhill crane, bald eagle,
osprey, beaver, river otter, muskrat
Major benthic insects: caddisflies
(*Brachycentrus*, *Glossosoma*,
Arctopsyche, *Cheumatopsyche*,
Hydropsyche), mayflies (*Baetis*,
Drunella, *Ephemerella*, *Rhithrogena*,
Tricorythodes), stoneflies (*Sweltsa*,
Zapada, *Claassenia*, *Hesperoperla*,
Skwala), true flies (chironomid
midges, black flies, *Tipula*)

Nonnative species: smallmouth bass,
walleye, largemouth bass, carp, brown
bullhead, black crappie, mosquitofish,
pumpkinseed, yellow perch, brown
trout, brook trout, channel catfish,
black bullhead, bluegill
Major riparian plants: black
cottonwood, willow, alder
Special features: 5 expansive floodplain
segments; headwaters in Alpine Lakes
Wilderness; rich agriculture area, within
Yakama Indian reservation
Fragmentation: 6 headwater storage
reservoirs; 7 irrigation diversion dams

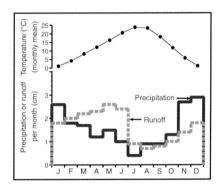

Yakima River in Yakima Canyon below Ellensburg, Washington (photo by Tim Palmer).

Willamette River

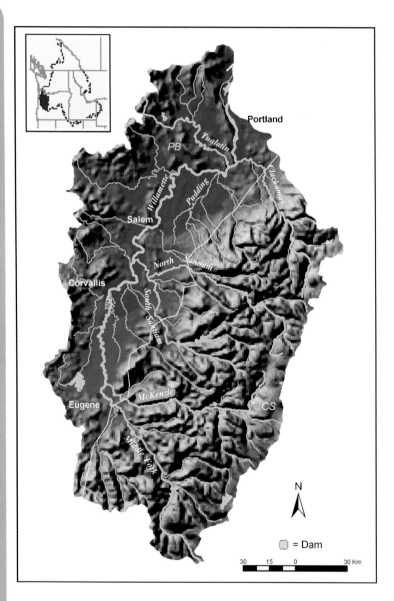

Relief: 3048 m
Basin area: 29,728 km^2
Mean discharge: 917 m^3/s
Mean annual precipitation: 153 cm

Mean air temperature: 11.9°C
Mean water temperature: 13.3°C
No. of fish species: 61 (~31 native)
No. of endangered species: 5

Physiographic provinces: Cascade–Sierra Mountains (CS), Pacific Border (PB)

Major fishes: mountain whitefish, northern pikeminnow, largescale sucker, mountain sucker, white sturgeon, several lampreys, Oregon chub, coho salmon, sockeye salmon, rainbow trout, chiselmouth, peamouth chub, speckled dace, redside shiner, bridgelip sucker, threespine stickleback, mottled sculpin, torrent sculpin

Major other aquatic vertebrates: bullfrog, western pond turtle, spotted frog, painted turtle, clouded salamander, bald eagle, great blue heron, beaver

Major benthic insects: caddisflies (*Hydropsyche, Cheumatopsyche, Lepidostoma, Heteroplectron, Glossosoma, Dicosmoecus*), stoneflies (*Taeniopteryx, Nemoura, Yoraperla*), mayflies (*Rhithrogena, Baetis, Paraleptophlebia, Ephemerella*), true flies (*Lipsothrix, Rheotanytarsus*)

Nonnative species: largemouth bass, smallmouth bass, bluegill, walleye, crappie, common carp, grass carp, brown bullhead, western mosquitofish, brook trout

Major riparian plants: black cottonwood, bigleaf maple, Oregon ash, Douglas fir

Special features: Willamette Valley is a major agriculture area; richest fish assemblage in Columbia basin

Fragmentation: 13 tributary dams regulate flow; 24 hydropower facilities

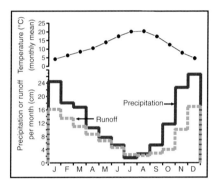

Willamette River at Oregon City, Oregon (photo by Tim Palmer).

Owyhee River

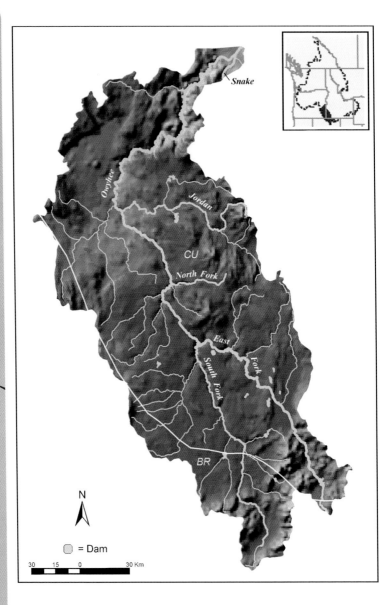

Relief: 2484 m
Basin area: 28,617 km^2
Mean discharge: 12 m^3/s
Mean annual precipitation: 31 cm

Mean air temperature: 10.6°C
Mean water temperature: 12.3°C
No. of fish species: 49 (25 native)
No. of endangered species: 0

Physiographic provinces: Columbia–Snake River Plateaus (CU), Basin and Range (BR)

Major fishes: yellow perch, white crappie, speckled dace, redside shiner, redband trout, sculpin, largescale sucker, flathead minnow

Major other aquatic vertebrates: Columbia spotted frog, Woodhouse's toad, bald eagle, white face ibis, leopard frog, bullfrog

Major benthic insects: caddisflies (*Hydropsyche*, *Cheumatopsyche*), stoneflies (*Taeniopteryx*, *Pteronarcys*), mayflies (*Rhithrogena*, *Baetis*, *Ephemerella*)

Nonnative species: black crappie, bluegill, brook trout, brown trout, black bullhead, brown bullhead, channel catfish, carp, flathead minnow, largemouth bass, Lahontan tui chub, oriental weatherfish, pumpkinseed, rainbow trout, smallmouth bass, tadpole madtom

Major riparian plants: black cottonwood, juniper, red-osier dogwood, alder, willow

Special features: driest subbasin in Columbia basin; only 16% of stream network flows year-round; some segments of main stem and tributaries are National Wild and Scenic Rivers

Fragmentation: Owyhee Dam near mouth; many irrigation diversions

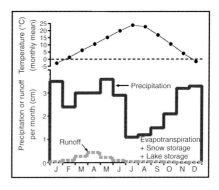

Owyhee River (photo from Momentum River Expeditions)

Grande Ronde River

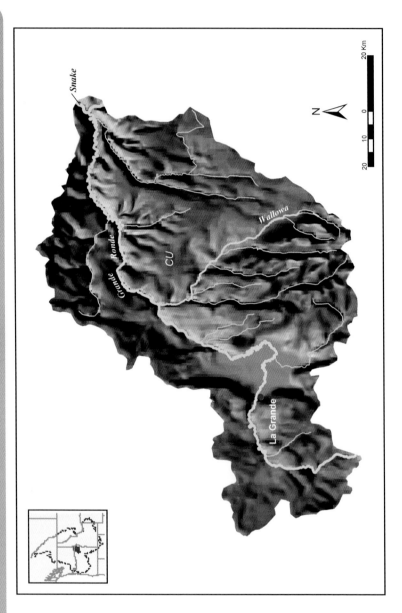

Relief: 2042 m

Basin area: 10,360 km^2

Mean discharge: 88 m^3/s

Mean annual precipitation: 46 cm

Mean air temperature: 7.6°C

Mean water temperature: 11.8°C

No. of fish species: 38 (23 native)

No. of endangered species: 4

Physiographic province: Columbia–
Snake River Plateaus

Major fishes: chinook salmon,
steelhead, bull trout, mountain
whitefish, brook trout, sculpin,
northern pikeminnow, peamouth chub,
longnose dace, speckled dace, redside
shiner, largescale sucker, bridgelip
sucker, mountain sucker, redband trout

Major other aquatic vertebrates: boreal
toad, painted turtle, tailed frog, chorus
frog, beaver, muskrat, mink, river otter,
great blue heron, wood duck

Major benthic insects: caddisflies
(*Hydropsyche*, *Cheumatopsyche*,
Glossosoma, *Rhyacophila*),
stoneflies (*Taeniopteryx*,
Hesperoperla), mayflies
(*Baetis*, *Ephemerella*)

Nonnative species: brook
trout, carp, lake trout,
bluegill, pumpkinseed,
warmouth, yellow perch,
black crappie, white
crappie, largemouth
bass, smallmouth bass,
channel catfish, flathead
catfish, brown bullhead

Major riparian plants:
black cottonwood,
Englemann spruce,
red-osier dogwood,
alder, willow, yellow
star thistle, diffuse
knapweed, spotted
knapweed

Special features:
National Wild and
Scenic designation for

portions of 4 tributaries and the lower
Grande Ronde

Fragmentation: two hydropower dams

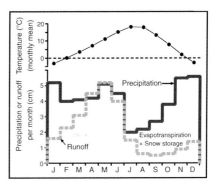

Grande Ronde River just upstream from its junction with Snake
River (photo by C. E. Cushing).

Clearwater River

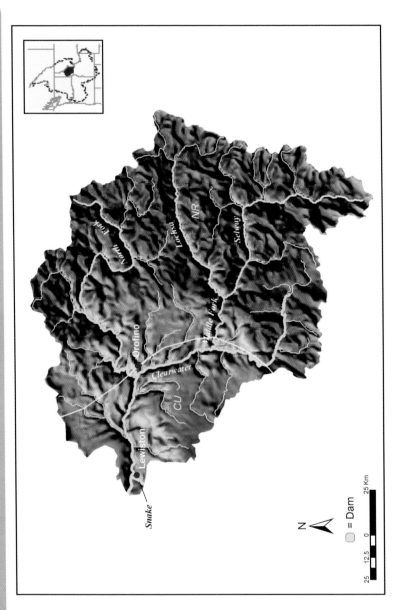

Relief: 2353 m   Mean air temperature: 11.5°C
Basin area: 31,080 km²   Mean water temperature: 9.5°C
Mean discharge: 433 m³/s   No. of fish species: 30 (19 native)
Mean annual precipitation: 74 cm   No. of endangered species: 2

Physiographic provinces: Northern Rocky Mountains (NR), Columbia–Snake River Plateaus (CU)

Major fishes: steelhead, chinook salmon, westslope cutthroat trout, brook trout, mountain whitefish, northern pikeminnow, chiselmouth, peamouth chub, longnose dace, speckled dace, redside shiner, largescale sucker, sculpin

Major other aquatic vertebrates: bullfrog, northern leopard frog, painted turtle, muskrat, beaver, river otter

Major benthic insects: caddisflies (*Arctopsyche*, *Hydropsyche*, *Cheumatopsyche*, *Glossosoma*, *Brachycentrus*), stoneflies (*Taeniopteryx*, *Pteronarcys*, *Hesperoperla*, *Claassenia*), mayflies (*Rhithrogena*, *Baetis*, *Ephemerella*), true flies (chironomid midges, black flies, *Hexatoma*, *Atherix*)

Nonnative species: kokanee, brook trout, golden trout, arctic grayling, tiger muskie, carp, channel catfish, brown bullhead, black bullhead, smallmouth bass, largemouth bass, bluegill, pumpkinseed, black crappie

Major riparian plants: black cottonwood, red-osier dogwood, alder, willows, sedges

Special features: major tributaries, Lochsa and Selway, are free flowing; Clearwater and tributaries mostly constrained in deep canyons; headwaters mostly designated wilderness

Fragmentation: Dworshak Dam on North Fork a major hydroelectric and storage facility

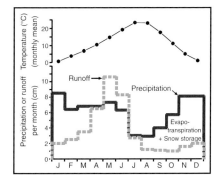

North Fork of the Clearwater River, Idaho (photo by C. E. Cushing).

Spokane River

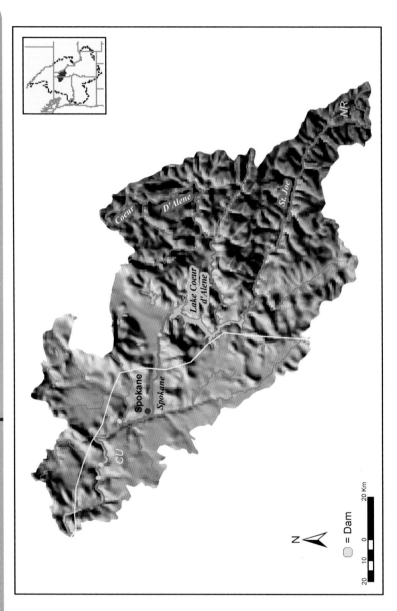

Relief: 1681 m
Basin area: 15,590 km²
Mean discharge: 225 m³/s
Mean annual precipitation: 63 cm

Mean air temperature: 7.5°C
Mean water temperature: 12.2°C
No. of fish species: 24
No. of endangered species: 0

Physiographic provinces: Northern Rocky Mountains (NR), Columbia–Snake River Plateaus (CU)

Major fishes: largemouth bass, yellow perch, tench, brown trout, largescale sucker, redside shiner, northern pikeminnow, chiselmouth, kokanee, westslope cutthroat trout, bull trout

Major other aquatic vertebrates: Columbia spotted frog, beaver, muskrat, white pelican, common loon, bald eagle

Major benthic insects: caddisflies (*Hydropsyche*, *Cheumatopsyche*, *Glossosoma*), stoneflies (*Taeniopteryx*), mayflies (*Rhithrogena*, *Baetis*, *Ephemerella*)

Nonnative species: rainbow trout, largemouth bass, yellow perch, tench, brown trout

Major riparian plants: black cottonwood, red-osier dogwood, willow

Special features: St. Joe River, one of the longest free-flowing rivers in Columbia basin; Lake Coeur d'Alene, a large glacial lake on main stem; Palouse Prairie, an expansive intermountain grassland, mostly cultivated

Fragmentation: Six low-head diversion and hydropower dams

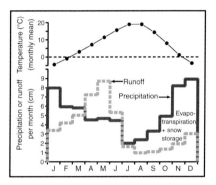

Spokane River below sewage plant near Spokane, Washington (photo by Tim Palmer).

Methow River

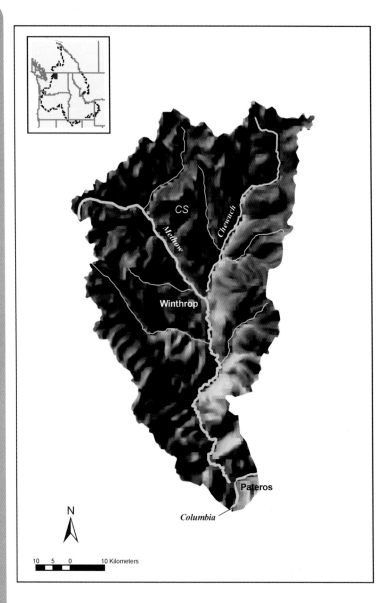

Relief: 2347 m
Basin area: 4831 km²
Mean discharge: 45 m³/s
Mean annual precipitation: 34 cm

Mean air temperature: 8.6°C
Mean water temperature: 9.5°C
No. of fish species: 32 (25 native)
No. of endangered species: 3

Physiographic province: Cascade–Sierra Mountains (CS)

Major fishes: chinook salmon, steelhead, bull trout, westslope cutthroat trout, bluegill, carp, bass, chiselmouth, brook trout, sculpin, largescale sucker, redside shiner, Pacific lamprey

Major other aquatic vertebrates: Columbia spotted frog, bald eagle, great blue heron, wood duck, mink

Major benthic insects: caddisflies (*Hydropsyche*, *Cheumatopsyche*, *Glossosoma*), stoneflies (*Claassenia*, *Hesperoperla*, *Isocapnia*, *Taeniopteryx*), mayflies (*Rhithrogena*, *Baetis*, *Ephemerella*)

Nonnative species: smallmouth bass, largemouth bass, bluegill, carp, brook trout, brown bullhead, black crappie

Major riparian plants: black cottonwood, Englemann spruce, red-osier dogwood, alder, willow

Special features: free flowing; headwaters in wilderness and North Cascades National Park; upper river with expansive floodplain but dry reach at low flow; lower river constrained in canyon; 2 of last (endangered) wild spring chinook salmon and steelhead runs in Columbia basin

Fragmentation: 7 small irrigation-diversion dams

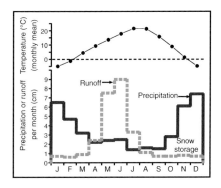

Lower Methow River, Washington (photo by Tim Palmer).

John Day River

Relief: 2682 m
Basin area: 20,980 km²
Mean discharge: 60 m³/s
Mean annual precipitation: 37 cm

Mean air temperature: 9.2°C
Mean water temperature: 12.5°C
No. of fish species: 27 (17 native)
No. of endangered species: 2

Physiographic province: Columbia–
Snake River Plateaus (CU)
Major fishes: chinook salmon, bull
trout, westslope cutthroat, mountain
whitefish, speckled dace, longnose
dace, redside shiner, chiselmouth,
largescale sucker, northern
pikeminnow, lamprey, sculpin
Major other aquatic vertebrates:
northern leopard frog, spotted frog,
western pond turtle, bullfrog, tailed
frog, painted turtle, beaver, muskrat
Major benthic insects: caddisflies
(*Hydropsyche*, *Cheumatopsyche*,
Glossosoma, *Brachycentrus*,
Cryptochia), stoneflies (*Taeniopteryx*,
Pteronarcys, *Hesperoperla*,
Claassenia), mayflies (*Rhithrogena*,
Baetis, *Ephemerella*)
Nonnative species: brook trout, carp,
black bullhead, brown bullhead, channel
catfish, largemouth bass, smallmouth
bass, black crappie, bluegill

Major riparian plants: black
cottonwood, red-osier dogwood, alder,
willow
Special features: unregulated river;
three segments designated National
Wild and Scenic Rivers; spring
chinook salmon and summer steelhead
populations are two of last remaining
intact wild populations of anadromous
fishes in Columbia River basin; lower
river very arid landscape
Fragmentation: irrigation diversions

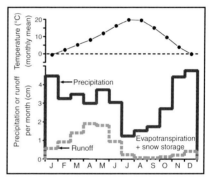

Lower John Day River, at Highway 206, Oregon (photo by A. C. Benke).

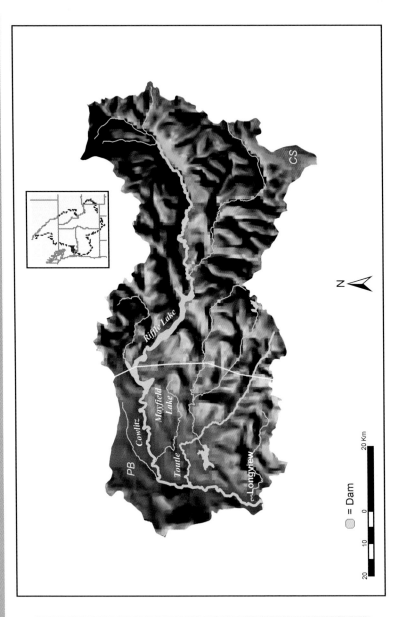

Cowlitz River

Relief: 4346 m
Basin area: 8870 km^2
Mean discharge: 261 m^3/s
Mean annual precipitation: 164 cm

Mean air temperature: 10.9°C
Mean water temperature: 10.4°C
No. of fish species: 32
No. of endangered species: 2

Physiographic provinces: Cascade–
Sierra Mountain (CS), Pacific Border
(PB)

Fragmentation: 3 hydroelectric dams
on main stem

Major fishes: chinook salmon, white
sturgeon, green sturgeon, pacific
lamprey, rainbow trout, largescale
sucker, bridgelip sucker, mountain
sucker, mountain whitefish, sculpin,
longnose dace, speckled dace, western
brook lamprey, northern pikeminnow,
brook trout

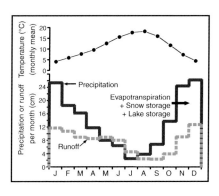

Major other aquatic vertebrates:
northwestern pond turtle, mink, blue
heron, bald eagle

Major benthic insects:
caddisflies (*Hydropsyche,
Cheumatopsyche,
Glossosoma*), stoneflies
(*Taeniopteryx, Pteronarcys,
Hesperoperla, Claassenia*),
mayflies (*Rhithrogena,
Baetis, Ephemerella*)

Nonnative species:
largemouth bass,
smallmouth bass, brook
trout, crappie, brown
bullhead, tiger
muskie

Major riparian plants: black
cottonwood, red-osier
dogwood, alder, willow

Special features: drains
highest point in Columbia
basin (Mount Rainier); river
choked with sediments after
1980 Mount St. Helens
volcanic eruption but
gradually recovering

Cowlitz River at Randle, Washington (photo by
Tim Palmer).

Chapter 14

Great Basin Rivers

Dennis K. Shiozawa and Russell B. Rader

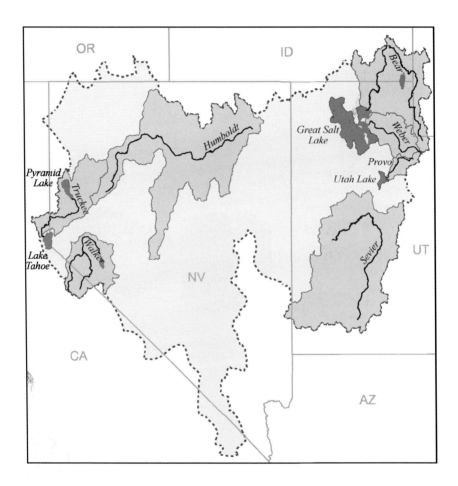

The Great Basin is the series of contiguous closed desert basins and mountain ranges that lie between the Sierra Nevada and southern Cascade Range mountains to the west, and the Wasatch Range and Plateau on the east. Most of this region is a cold desert, below freezing in the winter and hot in the summer. Nevada and Utah, which encompass the bulk of the Great Basin, are the two driest states in the United States. Thus Great Basin rivers are small with low discharge. Rivers of the Great Basin would be considered streams elsewhere, but the basins' dryness has enhanced their significance and they are as regionally important as are larger rivers in other parts of North America. Rivers and streams in the Great Basin often flow into valley playas, but the two largest subbasins with the Great Basin, the Bonneville and the Lahontan, have permanent terminal lakes.

Evidence of human habitation in the Great Basin dates to over 11,000 years ago. The earliest culture lived in caves around lakes and used a fluted point, similar to the Clovis point. Wetlands in the Carson and Owens valleys supported high human population densities and became the center of the Numic culture about 4500 years ago.[1] About 1000 years ago, the Numic people expanded into the northern and eastern Great Basin. The Fremont occupied the Bonneville Basin from 400 to 1350 A.D. but vanished about the time the Numic tribes are thought to have reached the eastern Great Basin. By the 1700s, the Great Basin was predominantly inhabited by Numic tribes: the Northern Paiute in the Lava Lakes Area and Lahontan Basin; the Mono, the Panamint, and the Kawaiisu in the southwestern Great Basin; the Shoshone in the Central Great Basin and upper Bonneville Basin; the Utes in the southern Bonneville Basin. The Spanish explored the Great Basin in the 1770s and fur trappers used the northern Great Basin in the early 1800s. In 1847 the Mormons settled the Bonneville Basin and established small communities throughout much of the Great Basin.

In this chapter, we describe the Bear, Sevier, Provo, and Weber rivers from the Bonneville basin and the Humboldt, Truckee, and Walker rivers from the Lahontan basin (see map). The largest of these is the Bear, with a virgin discharge of $71 \text{ m}^3/\text{s}$, which has been reduced to $52 \text{ m}^3/\text{s}$ today. The two largest basins, the Sevier ($>42,000 \text{ km}^2$) and the Humboldt ($>43,000 \text{ km}^2$) have seen far greater reductions in their flows (to only 4%) due to diversions.

---

1 Kelly, R. L. 2001. Prehistory of the Carson desert and Stillwater mountains. University of Utah Anthropological Papers. No. 123. University of Utah Press, Salt Lake City.

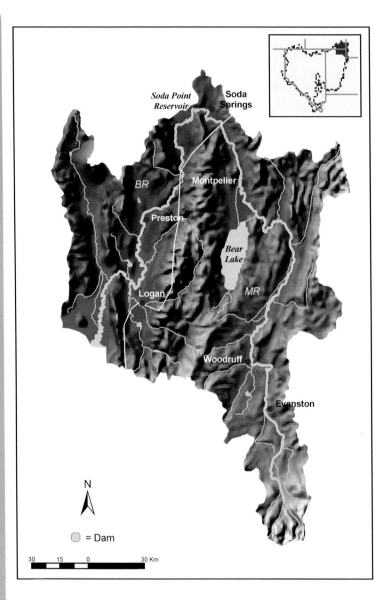

Soda Point Reservoir

Soda Springs

BR

Montpelier

Preston

Bear Lake

Logan

MR

Woodruff

Evanston

N

◯ = Dam

30   15   0   30 Km

Bear River

Relief: 2593 m
Basin area: 19,631 km²
Mean discharge: 71.2 m³/s
(virgin); 52.0 m³/s (present)
Mean annual precipitation: 56 cm

Mean air temperature: 6.5°C
Mean water temperature: 10.1°C
No. of fish species: 33
(17 native)
No. of endangered species: 0

Physiographic provinces: Middle Rocky Mountains (MR), Basin and Range (BR)

Major fishes: Bonneville cutthroat trout, mountain whitefish, Bonneville whitefish, Bear Lake whitefish, Bonneville cisco, speckled dace, longnose dace, redside shiner, leatherside chub, Utah chub, Utah sucker, mountain sucker, bluehead sucker, Paiute sculpin, mottled sculpin, Bear Lake sculpin

Major other aquatic vertebrates: Columbia spotted frog, boreal toad, Woodhouse's toad, northern leopard frog, tiger salamander, common garter snake, muskrat, beaver, dipper

Major benthic insects: mayflies (*Baetis*, *Drunella*, *Ephemerella*, *Epeorus*, *Rhithrogena*, *Stenonema*), stoneflies (*Sweltsa*, *Claassenia*, *Isoperla*, *Zapada*, *Skwala*), caddisflies (*Brachycentrus*, *Micrasema*, *Helicopsyche*, *Arctopsyche*, *Cheumatopsyche*, *Hydropsyche*, *Hydroptila*, *Lepidostoma*, *Nectopsyche*, *Oecetis*, *Chyranda*, *Dicosmoecus*, *Hesperophylax*, *Limnephilus*, *Polycentropus*, *Rhyacophila*, *Oligophlebodes*)

Nonnative species: common carp, brown trout, rainbow trout, brook trout, lake trout, Yellowstone cutthroat trout, sockeye salmon, black bullhead, channel catfish,

bluegill, green sunfish, yellow perch, walleye, largemouth bass, western mosquitofish, fathead minnow

Major riparian plants: Fremont cottonwood, narrowleaf cottonwood, river birch, red-osier dogwood, willow, box elder, wild rose

Special features: Bear Lake, Bear River Bay Bird Refuge, Uinta Mountains, Thatcher basin, Red Rock Pass, Oneida narrows

Fragmentation: 6 dams on main stem

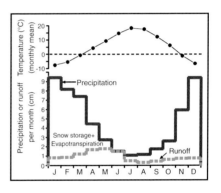

Bear River, cutting through lava flows near Grace, Idaho (photo by D. K. Shiozawa).

Sevier River

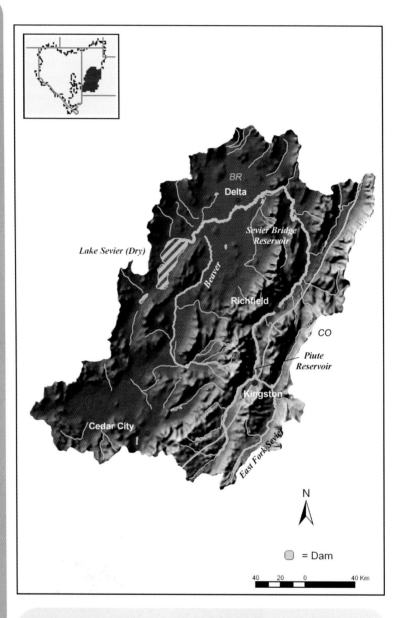

N

○ = Dam

40   20   0   40 Km

Relief: 2341 m
Basin area: 42,025 km^2
Mean discharge: 32.2 m^3/s
(virgin); 1.3 m^3/s (present)
Mean annual precipitation: 36 cm

Mean air temperature: 9.3°C
Mean water temperature: 11.3°C
No. of fish species: 21
(9 native)
No. of endangered species: 0

Physiographic provinces: Colorado
Plateaus (CO), Basin and Range (BR)
Major fishes: Bonneville cutthroat
trout, speckled dace, redside shiner,
leatherside chub, least chub, Utah
chub, mountain sucker, Utah sucker,
mottled sculpin
Major other aquatic vertebrates:
northern leopard frog, tiger
salamander, common garter snake,
muskrat, beaver
Major benthic insects: mayflies
(*Baetis, Rhithrogena*), stoneflies
(*Capnia, Utacapnia, Alloperla,
Claassenia, Hesperoperla, Diura,
Isogenoides, Isoperla, Megarcys*),
caddisflies (*Brachycentrus,
Hydropsyche, Hydroptila,
Ochrotrichia, Nectopsyche, Oecetis,
Amphicosmoecus, Hesperophylax,
Limnephilus, Onocosmoecus,
Rhyacophila, Oligophlebodes*)
Nonnative species: Asiatic clam,
common carp, brown trout, rainbow
trout, brook trout, Yellowstone

cutthroat trout, mountain whitefish,
channel catfish, black bullhead,
walleye, yellow perch, fathead
minnow, western mosquitofish,
tamarisk, Russian olive
Major riparian plants: Fremont
cottonwood, narrowleaf cottonwood,
river birch, red-osier dogwood, willow,
box elder, wild rose
Special features: desert river; usually
dries before reaching Sevier Lake;
National Parks in high plateaus
Fragmentation: 8 dams on main stem

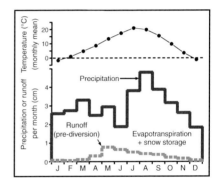

Sevier River at Big Rock Candy Mountain, Utah (photo by D. K. Shiozawa).

Humboldt River

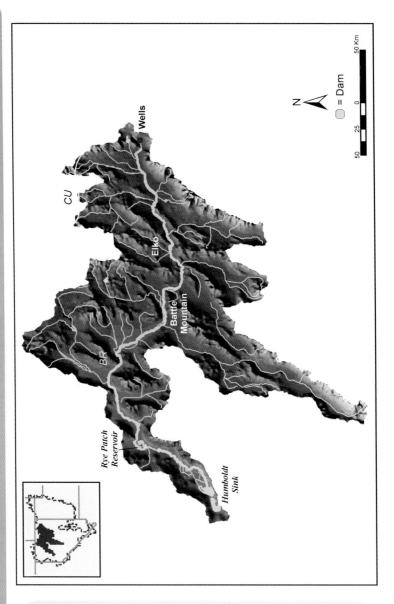

Relief: 2036 m
Basin area: 43,597 km^2
Mean discharge: 25.8 m^3/s
(virgin); 0.9 m^3/s (present)
Mean annual precipitation: 22 cm

Mean air temperature: 9.4°C
Mean water temperature: 12.3°C
No. of fish species: 23
(7 native)
No. of endangered species: 1

Physiographic province: Basin and Range (BR)

Major fishes: Lahontan cutthroat trout, Paiute sculpin, tui chub, Lahontan redside, speckled dace, Tahoe sucker, mountain sucker

Major other aquatic vertebrates: Columbia spotted frog, northern leopard frog, Great Basin spadefoot toad, common garter snake, western aquatic garter snake, muskrat, beaver

Major benthic insects: mayflies (*Baetis, Acentrella, Camelobaetidius, Centroptilum, Baetisca, Ephemerella, Ephemera, Hexagenia, Heptagenia, Rhithrogena, Tricorythodes, Traverella, Ephoron*), stoneflies (*Isogenoides, Isoperla, Capnura, Taenionema, Acroneuria*), caddisflies (*Brachycentrus, Micrasema, Anagapetus, Cheumatopsyche, Hydropsyche, Hydroptila, Chyranda, Limnephilus, Nectopsyche, Rhyacophila*)

Nonnative species: common carp, goldfish, brook trout, brown trout, rainbow trout, Yellowstone cutthroat trout, black bullhead, channel catfish, walleye, bluegill, green sunfish, black crappie

Major riparian plants: quaking aspen, Fremont cottonwood, black cottonwood, narrowleaf cottonwood, willow, river birch

Special features: longest river in Great Basin; terminates in Humboldt Sink, once a large wetland

Fragmentation: 1 dam on main stem, numerous diversions in lower basin

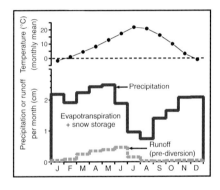

Humboldt River, east of Carlin, Nevada (photo by D. K. Shiozawa).

Truckee River

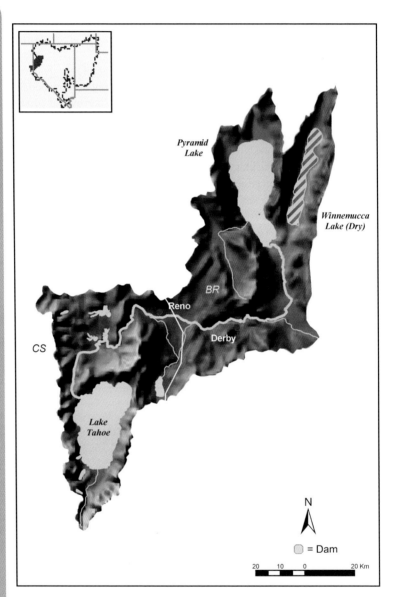

Pyramid
Lake

Winnemucca
Lake (Dry)

BR

Reno

Derby

CS

Lake
Tahoe

N

◯ = Dam

20   10   0   20 Km

Relief: 2159 m
Basin area: 7925 km²
Mean discharge: 18.6 m³/s
(virgin); 14.2 m³/s (present)
Mean annual precipitation: 43 cm

Mean air temperature: 8.4°C
Mean water temperature: 11.4°C
No. of fish species: 21
(8 native)
No. of endangered species: 3

Physiographic provinces: Cascade–Sierra Mountains (CS), Basin and Range (BR)

Major fishes: Lahontan cutthroat trout, Paiute sculpin, tui chub, Lahontan redside, speckled dace, Tahoe sucker, mountain sucker, cui-ui

Major other aquatic vertebrates: Pacific tree frog, mountain yellow-legged frog, northern leopard frog, river otter, muskrat, beaver

Major benthic insects: mayflies (*Ephemerella*, *Drunella*, *Epeorus*, *Heptagenia*, *Rhithrogena*, *Cinygmula*, *Baetis*), stoneflies (*Capnia*, *Utacapnia*, *Skwala*, *Sweltsa*, *Malenka*, *Kogotus*, *Paraleuctra*, *Perlomyia*, *Prostoia*, *Claassenia*, *Cultus*, *Isoperla*), caddisflies (*Glossosoma*, *Arctopsyche*, *Hydropsyche*, *Parapsyche*, *Lepidostoma*, *Rhyacophila*)

Nonnative species: common carp, brook trout, brown trout, rainbow trout, Yellowstone cutthroat trout, sockeye salmon, black bullhead, channel catfish, bluegill, green sunfish, Sacramento perch, smallmouth bass

Major riparian plants: mountain alder, quaking aspen, black cottonwood, Fremont cottonwood, narrowleaf cottonwood, willow, red-osier dogwood, river birch

Special features: desert river flowing through Lake Tahoe and terminating in Pyramid Lake; Lake Tahoe recreation area

Fragmentation: 4 hydroelectric diversion dams, one diversion dam in lower basin (Derby Dam), several small diversion dams on main stem

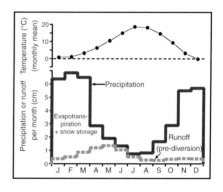

Truckee River near Reno, Nevada (photo by D. K. Shiozawa).

Provo River

Relief: 2294 m
Basin area: 1761 km^2
Mean discharge: 18.1 m^3/s
(virgin); 9.9 m^3/s (present)
Mean annual precipitation: 64 cm

Mean air temperature: 9.3°C
Mean water temperature: 7.7°C
No. of fish species: 28
(13 native)
No. of endangered species: 1

Physiographic provinces: Middle Rocky Mountains (MR), Basin and Range (BR)

Major fishes: Bonneville cutthroat trout, mountain whitefish, redside shiner, leatherside chub, Utah chub, speckled dace, longnose dace, mountain sucker, Utah sucker, June sucker, mottled sculpin, Paiute sculpin

Major other aquatic vertebrates: Columbia spotted frog, northern leopard frog, Pacific chorus frog, Woodhouse's toad, boreal toad, tiger salamander, beaver, muskrat, mink, river otter, water vole, dipper

Major benthic insects: mayflies (*Baetis, Paraleptophlebia, Drunella, Rhithrogena*), stoneflies (*Pteronarcys, Claassenia, Hesperoperla, Diura, Isogenoides, Isoperla, Megarcys*), caddisflies (*Brachycentrus, Hydropsyche, Hydroptila, Oecetis, Hesperophylax, Limnephilus, Onocosmoecus, Rhyacophila*)

Nonnative species: virile crayfish, common carp, brown trout, rainbow trout, brook trout, Yellowstone cutthroat trout, channel catfish, black bullhead, largemouth bass, white bass, yellow perch, western mosquitofish, tamarisk, Russian olive

Major riparian plants: Fremont cottonwood, narrowleaf cottonwood, river birch, red-osier dogwood, willow

Special features: one of top trout streams in North America

Fragmentation: 2 dams on main stem

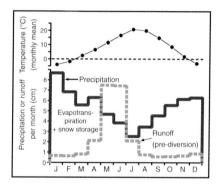

Provo River near Midway, Utah (photo by D. K. Shiozawa).

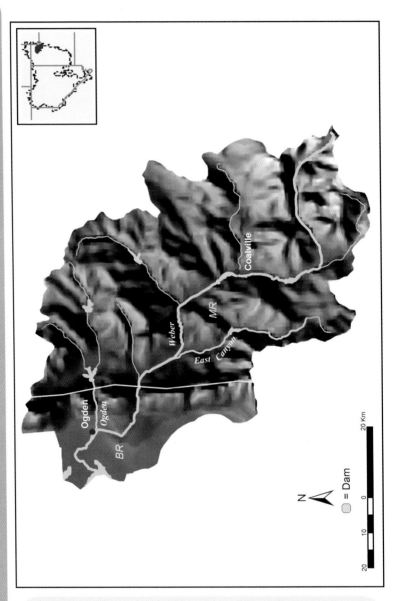

Weber River

Relief: 2358 m
Basin area: 6070 km^2
Mean discharge: 38.3 m^3/s
(virgin); 12.2 m^3/s (present)
Mean annual precipitation: 53 cm

Mean air temperature: 7.7°C
Mean water temperature: 10.0°C
No. of fish species: 26
(11 native)
No. of endangered species: 0

Physiographic provinces: Middle
Rocky Mountains (MR), Basin and
Range (BR)
Major fishes: redside shiner, Utah
chub, speckled dace, longnose dace,
mottled sculpin, Paiute sculpin,
Bonneville cutthroat trout, mountain
whitefish, mountain sucker, Utah
sucker, bluehead sucker
Major other aquatic vertebrates:
Woodhouse's toad, tiger salamander,
northern leopard frog, boreal toad,
Pacific chorus frog, Columbia spotted
frog, Great Basin spadefoot toad, water
vole, muskrat, beaver, mink, river otter
Major benthic insects: mayflies (*Baetis,
Rhithrogena, Drunella*), stoneflies
(*Capnia, Utacapnia, Alloperla,
Pteronarcys, Amphinemura, Podmosta,
Prostoia, Claassenia, Hesperoperla,
Diura, Isogenoides, Isoperla*), caddisflies
(*Brachycentrus, Hydropsyche,
Hydroptila, Hesperophylax,
Limnephilus, Oligophlebodes*)
Nonnative species: tamarisk, Russian
olive, common carp, rainbow trout,
Yellowstone cutthroat trout, brown

trout, brook trout, yellow perch,
smallmouth bass, largemouth bass,
black bullhead, channel catfish,
green sunfish, bluegill, black crappie,
walleye, western mosquitofish
Major riparian plants: Fremont
cottonwood, narrowleaf cottonwood,
river birch, red-osier dogwood,
willow, box elder
Special features: mountainous river
system with most of basin privately
owned; second-largest tributary of
Great Salt Lake
Fragmentation: 7 major dams, 2 on
main stem

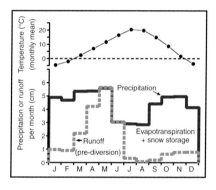

Weber River, west of Coalville, Utah (photo by D. K. Shiozawa).

Walker River

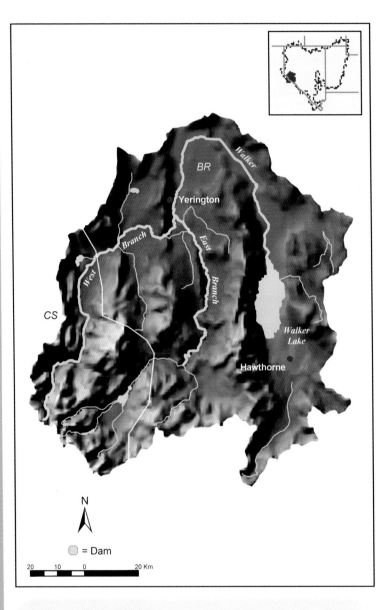

N

= Dam

20   10   0   20 Km

Relief: 2576 m
Basin area: 7894 km^2
Mean discharge: 9.7 m^3/s
(virgin); 3.5 m^3/s (present)
Mean annual precipitation: 38 cm

Mean air temperature: 23.7°C
Mean water temperature: 10.8°C
No. of fish species: 15
(7 native)
No. of endangered species: 2

Physiographic provinces: Cascade–Sierra Mountains (CS), Basin and Range (BR)

Major fishes: Lahontan redside, speckled dace, tui chub, Paiute sculpin, Lahontan cutthroat trout, mountain sucker, Tahoe sucker

Major other aquatic vertebrates: mountain yellowlegged frog, Yosemite toad, northern leopard frog, boreal toad, Great Basin spadefoot toad, beaver, muskrat, mink, common loon

Major benthic insects: mayflies (*Baetis, Heptagenia, Rhithrogena, Ephemerella*), stoneflies (*Capnia, Pteronarcys, Utacapnia, Sweltsa, Isoperla, Prostoia, Paraleuctra, Claassenia, Kogotus*), caddisflies (*Brachycentrus, Micrasema, Lepidostoma, Hydropsyche, Hydroptila, Nectopsyche, Limnephilus, Rhyacophila, Glossosoma*)

Nonnative species: Russian olive, common carp, channel catfish, black bullhead, Sacramento perch, white bass, brown trout, rainbow trout, Yellowstone cutthroat trout

Major riparian plants: narrowleaf cottonwood, Fremont cottonwood, river birch, red-osier dogwood, willow, mountain alder

Special features: best-studied basin in Lahontan basin; terminates in Walker Lake, which has major cutthroat trout fishery; lake faces destruction from

increased salinity due to diversions and groundwater pumping

Fragmentation: 9 major reservoirs, 4 on East Walker River, 4 on West Walker River, one on main Walker River

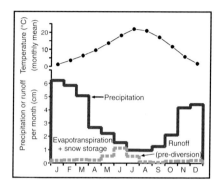

East Walker River at the Elbow, Lyon County, Nevada (photo by D. K. Shiozawa).

Chapter 15

Fraser River Basin

Trefor B. Reynoldson, Joseph Culp, Rick Lowell, and John S. Richardson

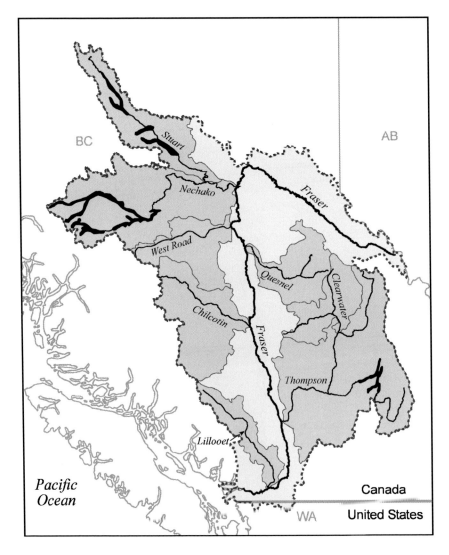

DOI: 10.1016/B978-0-12-375088-4.00015-4

The Fraser River, with its headwaters in the Rocky Mountains, flows across the dry Fraser Plateau through coastal mountain ranges to the Pacific Ocean (see map). By any measure the Fraser is one of Canada's great rivers with an abundance and diversity of natural resources that rival almost any other river in the world. Of Canadian rivers it has the third highest mean flow ($3972\,m^3/s$), is the fifth longest (1375 km), and has the fifth largest drainage basin ($234,000\,km^2$). The diversity of geology, climate, and the landscape is so great that it includes 11 of British Columbia's 14 biogeoclimatic zones before discharging into the Straits of Georgia in the City of Vancouver. Beyond the main-stem river is a vast network of tributaries that envelop more than a quarter of British Columbia. The entire catchment reaches as far north as Bulkley House on the Stuart River and stretches westwards from the highest summits of the Coast Mountain range to the heights of the Rocky Mountains in the east.

For at least 10,000 years native peoples have occupied the Fraser River Basin and used its resources.[1] Within British Columbia there is a greater degree of indigenous cultural and linguistic diversity than in any other region of Canada. The Fraser Basin was and remains inhabited by native peoples speaking six separate and distinct languages. Within the basin can be distinguished both Northwest-coast cultures on the lower part of the river and adjoining sea coast, and Plateau cultures in the middle and upper parts. Prior to contact with European explorers, an estimated 50,000 people lived in the Basin.[2] However, these numbers were likely already half the original population after the ravages of diseases from earlier explorers by the time Simon Fraser canoed the river some two hundred years later in 1808. The discovery of gold on the Thompson tributary in 1857 caused an influx of people raising the total population to pre-contact levels. The gold strikes in the headwaters of the Willow River resulted in the development of famous land and water routes, including the Harrison Trail and Cariboo Road up the Fraser and provided the impetus to use the river as a path through the coastal mountain barrier.

In this chapter, we describe the main-stem Fraser River as well as eight tributaries that illustrate the wide range of geologic, climatic, and biological diversity in the basin (see map). These tributaries include the Thompson, Nachako, Stuart, West Road, Quesnel, Chilcotin, Lillooet/Harrison, and Clearwater rivers, the largest of which is the Thompson with a mean discharge of $787\,m^3/s$.

---

1 Kew, J. E. M. & J. R. Griggs. 1991. Native Indians of the Fraser Basin: towards a model of sustainable resource use. In Dorcey, H. J. and J. R. Griggs (eds.). *Water in Sustainable Development: exploring our common future in the Fraser River basin.* Vol. 1. Westwater Research Centre, University of British Columbia, Vancouver.

2 Dorcey, A. H. J. 1991. Water in the sustainable development of the Fraser River Basin. In H. J. Dorcey and J. R. Griggs (eds.). *Water in Sustainable Development: exploring our common future in the Fraser River basin.* Vol. 2, pp. 3–18. Westwater Research Centre, University of British Columbia, Vancouver.

Fraser River

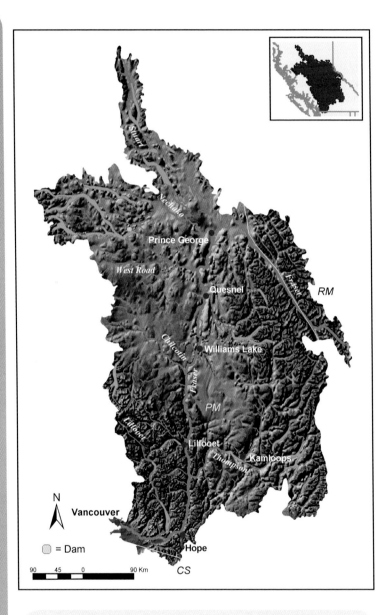

Relief: 3954 m
Basin area: 234,000 km²
Mean discharge: 3972 m³/s
Mean annual precipitation: 80 cm

Mean air temperature: 6.1°C
Mean water temperature: NA
No. of fish species: 40 fresh
water (native), 8 marine
No. of endangered species: 7

Physiographic Provinces: Coast Mountains of British Columbia and Southeast Alaska (PM), Rocky Mountains in Canada (RM)
Major fishes: starry flounder, coho salmon, chinook salmon, sockeye salmon, pink salmon, chum salmon, rainbow trout, river lamprey, pacific lamprey, eulachon, surf smelt, longfin smelt, northern squawfish, peamouth chub, redside shiner, longnose dace, largescale sucker, longnose sucker, white sucker, bridgelip sucker, prickly sculpin
Major other aquatic vertebrates: great blue heron, bald eagle, river otter, mink
Major benthic insects: mayflies (*Baetis*, *Ephemerella*, *Drunella*, *Rhithrogena*), stoneflies (*Capnia*, *Sweltsa*, *Taenionema*, *Zapada*), true flies (*Eukiefferiella*, *Micropsectra*, *Tvetenia*), caddisflies (*Rhyacophila*)
Nonnative species: American shad, common carp, brown bullhead, goldfish, fathead minnow, Atlantic salmon, brook

trout, yellow perch, pumpkinseed, largemouth bass, black crappie
Major riparian plants: white spruce, lodgepole pine, trembling aspen, Douglas fir, Engelmann spruce, alpine fir, common paper birch, black cottonwood, trembling aspen, various willows, various grasses
Special features: Canadian Heritage river; Fraser Canyon between Yale and Boston Bar
Fragmentation: none except for dam in Nechako River

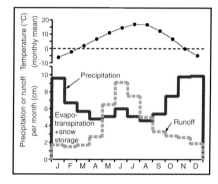

Fraser River at Hope, British Columbia (photo by Tim Palmer).

Thompson River

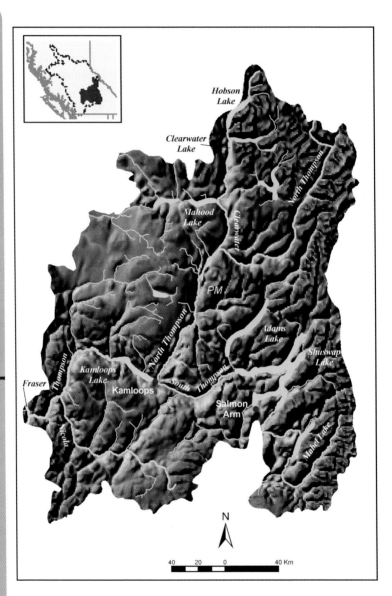

Relief: 3250 m
Basin area: 55,400 km²
Mean discharge: 787 m³/s
Mean annual precipitation: 43 cm

Mean air temperature: 9.7°C
Mean water temperature: 9.0°C
No. of fish species: 24
No. of endangered species: 2

Physiographic province: Coast
Mountains of British Columbia and
Southeast Alaska (PM)

Major fishes: round whitefish, coho
salmon, chinook salmon, sockeye
salmon, pink salmon, largescale
sucker, bridgelip sucker, northern
squawfish, longnose dace, slimy
sculpin

Major other aquatic vertebrates:
beaver, muskrat, dipper, merganser,
osprey

Major benthic insects:
mayflies (*Baetis*, *Ephemerella*,
Paraleptophlebia, *Rhithrogena*),
caddisflies
(*Arctopsyche*,
Brachycentrus,
Cheumatopsyche,
Glossosoma,
Hydropsyche,
Hydroptila),
stoneflies
(*Arcynopteryx*,
Skwala), true flies
(*Cardiocladius*,
Cricotopus,
Eukiefferiella)

Nonnative species:
carp

Major riparian
plants: common
paper birch, black
cottonwood,
trembling aspen,
various willows,
various grasses

Special features:
white-water rapids

between Spences Bridge and the
confluence with the Fraser River;
Adams River run of Sockeye salmon

Fragmentation: none

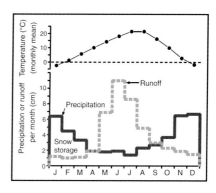

North Fork of the Thompson River above Vavenby, British Columbia
(photo by Tim Palmer).

Nechako River

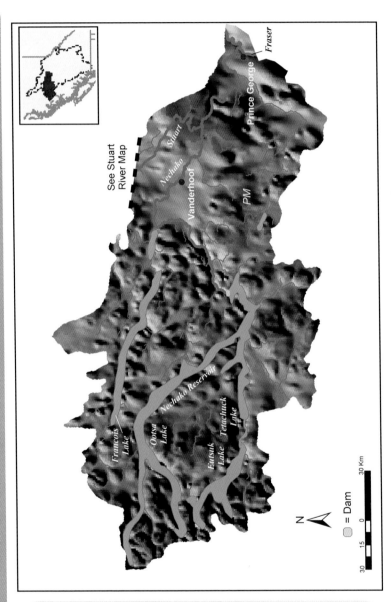

Relief: ~1400 m
Basin area: 42,500 km²
Mean discharge: 434 m³/s (virgin);
284 m³/s (present)
Mean annual precipitation: 60 cm

Mean air temperature: 3.7°C
Mean water temperature: NA
No. of fish species: 26
No. of endangered species: 2

Physiographic province: Coast Mountains of British Columbia and Southeast Alaska (PM)

Major fishes: burbot, mountain whitefish, lake whitefish, lake trout, chinook salmon, sockeye salmon (kokanee), rainbow trout, prickly sculpin, largescale sucker, longnose sucker, redside shiner, peamouth chub, lake chub

Major other aquatic vertebrates: beaver, moose, merganser, goldeneye duck

Major benthic insects: mayflies (Baetidae, *Ephemerella*, *Serratella*), stoneflies (Capniidae, Chloroperlidae, *Hesperoperla pacifica*, *Pteronarcys*), caddisflies (Hydropsychidae), chironomid midges (Tanypodinae, Tanytarsini, Orthocladiinae)

Nonnative species: goldfish, brook trout

Major riparian plants: willows, black cottonwood, balsam poplar, alder, aspen

Special features: remote basin with large areas of pristine forest in the western part of its basin near the Coast Range

Fragmentation: major dam with large influence on river regulation and diversion of water to a power generating station on the Pacific coast

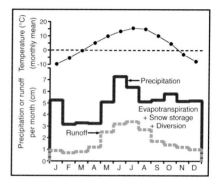

Nechako River at Highway 16, British Columbia (photo by Tim Palmer).

Stuart River

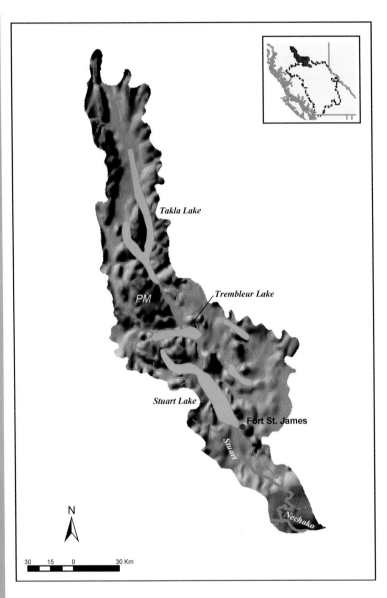

Takla Lake

Trembleur Lake

PM

Stuart Lake

Fort St. James

Stuart

Nechako

N

30   15   0   30 Km

Relief: 1097 m   Mean air temperature: 2.8°C
Basin area: 14,600 km^2 Mean water temperature: 6.8°C
Mean discharge: 128 m^3/s   No. of fish species: 23
Mean annual precipitation: 49 cm   No. of endangered species: 2

Physiographic province: Coast Mountains of British Columbia and Southeast Alaska (PM)

Major fishes: mountain whitefish, lake whitefish, lake trout, sockeye salmon, rainbow trout, largescale sucker, slimy sculpin, northern pikeminnow, redside shiner, peamouth chub, lake chub

Major other aquatic vertebrates: beaver, river otter, muskrat, moose, merganser

Major benthic insects: mayflies (*Diphetor, Ephemerella, Leucrocuta, Rhithrogena*), stoneflies (*Capnia, Paraleuctra, Sweltsa, Taenionema*), true flies (*Micropsectra, Polypedilum, Rheocricotopus, Tanytarsus, Tvetenia*)

Nonnative species: none

Major riparian plants: black cottonwood, Sitka alder, willows

Special features: lake-dominated system with short tributary streams flowing into Takla and Stuart lakes; system supports sockeye salmon populations with longest migration in Fraser River system

Fragmentation: none

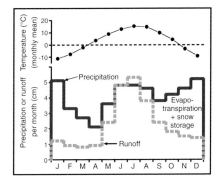

Stuart River, British Columbia (photo by J. Burrows).

West Road River

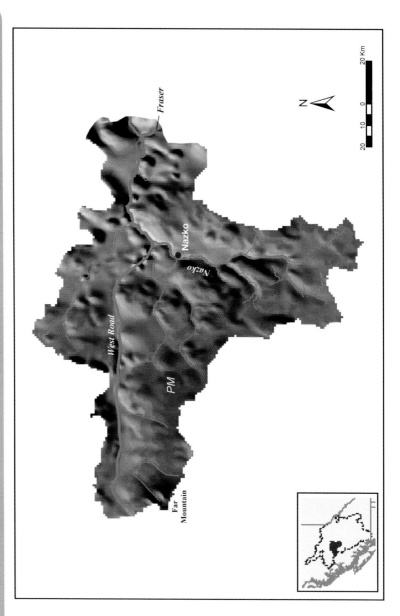

Relief: 2000 m
Basin area: 12,400 km^2
Mean discharge: 38 m^3/s
Mean annual precipitation: 64 cm

Mean air temperature: 4.7°C
Mean water temperature: NA
No. of fish species: 21
No. of endangered species: 0

Physiographic province: Coast Mountains of British Columbia and Southeast Alaska (PM)

Major fishes: burbot, mountain whitefish, bull trout, Chinook salmon, sockeye salmon, rainbow trout, largescale sucker, longnose sucker, redside shiner, northern squawfish, peamouth chub

Major other aquatic vertebrates: great blue heron, bald eagle, river otter, mink

Major benthic insects: mayflies (*Baetis, Paraleptophlebia, Leucrocuta, Serratella*), stoneflies (*Capnia, Sweltsa, Zapada*), true flies (*Micropsectra, Polypedilum, Tanytarsus, Tvetenia, Zavrelimyia*)

Nonnative species: brook trout

Major riparian plants: willows, black cottonwood, alder, aspen

Special features: one of the finest trout streams in British Columbia; route chosen by Sir Alexander Mackenzie in 1793—now a heritage hiking trail to Bella Coola

Fragmentation: none

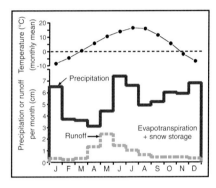

West Road River, British Columbia (photo by J. Burrows).

Quesnel River

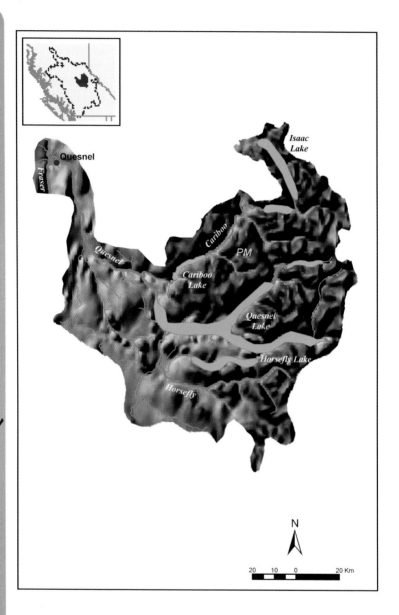

Relief: 2500 m

Basin area: 11,500 km²

Mean discharge: 239 m³/s

Mean annual precipitation: 54 cm

Mean air temperature: 4.9°C

Mean water temperature: NA

No. of fish species: 23

No. of endangered species: 0

Physiographic province: Coast
Mountains of British Columbia and
Southeast Alaska (PM)

Major fishes: sockeye salmon, rainbow
trout, largescale sucker, longnose
sucker, redside shiner, northern
squawfish, peamouth chub, lake chub

Major other aquatic vertebrates:
moose, beaver, merganser

Major benthic insects: NA

Nonnative species: none

Major riparian plants: white spruce,
lodgepole pine, trembling aspen,
Douglas fir

Special features: remarkable comeback
of Quesnel River sockeye; run of

12.2 million fish in 1993, with
2.5 million fish making it to Quesnel
headwaters to spawn; once again the
Fraser's greatest sockeye producer,
surpassing world-famous Adams River
for sockeye production

Fragmentation: none

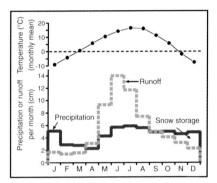

Quesnel River, British Columbia (photo by R. Holmes).

Chilcotin River

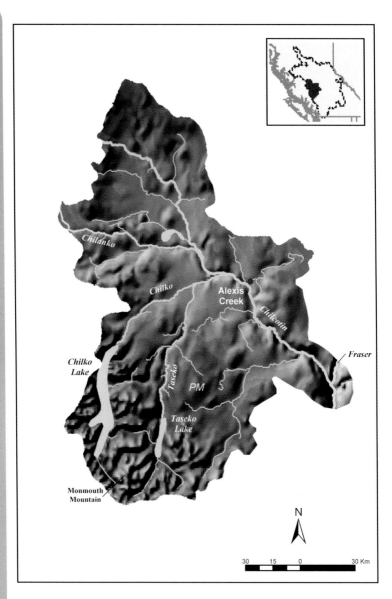

Relief: 3250 m
Basin area: 19,300 km²
Mean discharge: 102 m³/s
Mean annual precipitation: 34 cm

Mean air temperature: 2.2°C
Mean water temperature: NA
No. of fish species: 16
No. of endangered species: 0

Physiographic province: Coast Mountains of British Columbia and Southeast Alaska (PM)

Major fishes: mountain whitefish, rainbow trout, sockeye salmon, bull trout, longnose sucker, redside shiner, largescale sucker, northern squawfish, peamouth chub, lake chub

Major other aquatic vertebrates: beaver, moose, merganser

Major benthic insects: mayflies (*Baetis*, *Epeorus*, *Paraleptophlebia*, *Rhithrogena*), stoneflies (*Capnia*, *Sweltsa*, *Taenionema*, *Zapada*), true flies (*Brillia*, *Eukiefferiella*, *Micropsectra*, *Orthocladius*)

Nonnative species: brook trout

Major riparian plants: white spruce, lodgepole pine, trembling aspen,

Douglas fir, Engelmann spruce, alpine fir, bunchgrass

Special features: among best and most challenging river in North America for kayaking and white-water rafting; spectacular scenery, such as Farwell Canyon, where river cuts deeply into sandstone cliffs with native pictographs on overhang

Fragmentation: none

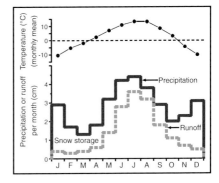

Chilcotin River at Farwell Canyon, British Columbia (photo by M. Hernandez-Henriquez).

Clearwater River

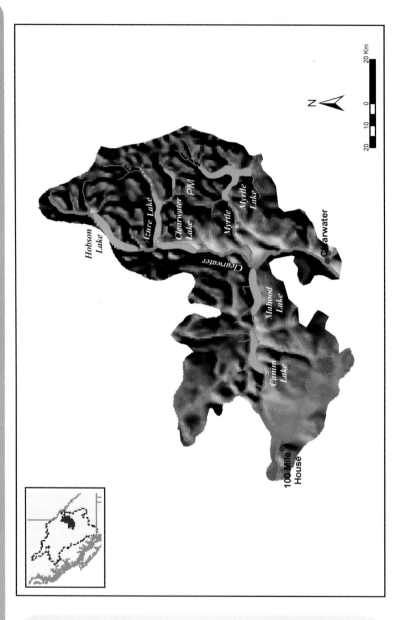

Relief: 2750 m
Basin area: 10,200 km²
Mean discharge: 223 m³/s
Mean annual precipitation: 45 cm

Mean air temperature: 6.2°C
Mean water temperature: NA
No. of fish species: 16
No. of endangered species: 0

Physiographic province: Coast Mountains of British Columbia and Southeast Alaska (PM)

Major fishes: mountain whitefish, chinook salmon, sockeye salmon, coho salmon, rainbow trout, largescale sucker, longnose sucker

Major other aquatic vertebrates: beaver, moose, merganser

Major benthic insects: mayflies (*Baetis, Drunella, Ephemerella, Paraleptophlebia, Rhithrogena*), stoneflies (*Sweltsa, Zapada*), true flies (*Micropsectra, Tvetenia, Chelifera*)

Nonnative species: none

Major riparian plants: white spruce, lodgepole pine, trembling aspen, Douglas fir, Engelmann spruce, alpine fir

Special features: headwaters rise in Wells Gray Provincial Park; 6 major lakes; numerous spectacular waterfalls; incredible mountain scenery and hiking trails; Helmcken Falls plunges 141 m down narrow canyon

Fragmentation: none

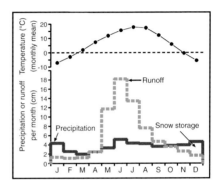

Clearwater River, British Columbia (photo by M. Taylor).

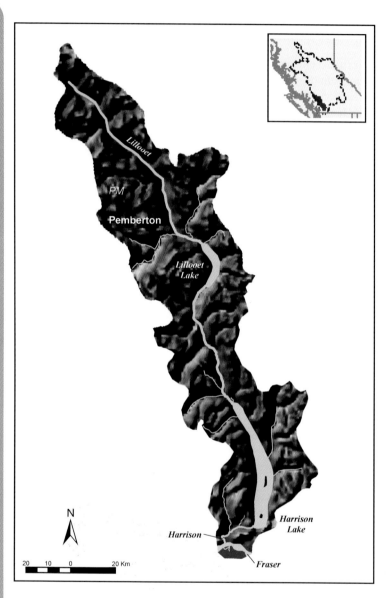

Relief: 2750 m
Basin area: 7870 km²
Mean discharge: 445 m³/s
Mean annual precipitation: 173 cm

Mean air temperature: 10.2°C
Mean water temperature: NA
No. of fish species: 20
No. of endangered species: 0

Physiographic province: Coast Mountains of British Columbia and Southeast Alaska (PM)

Major fishes: chinook salmon, sockeye salmon, coho salmon, chum salmon, pink salmon, rainbow trout, cutthroat trout, largescale sucker, redside shiner

Major other aquatic vertebrates: beaver, moose, merganser

Major benthic insects: mayflies (*Drunella*, *Rhithrogena*), stoneflies (*Megarcys*, *Taenionema*), caddisflies (*Oligophlebodes*)

Nonnative species: brown bullhead

Major riparian plants: white spruce, lodgepole pine, trembling aspen, Douglas fir, Engelmann spruce, alpine fir

Special features: headwater Lillooet River runs for almost 200 km of white-water paddling with spectacular autumn sockeye salmon runs; Lillooet empties into Harrison Lake, the largest lake in southwestern British Columbia

Fragmentation: none

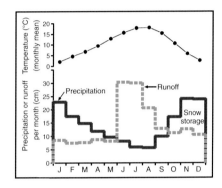

Lillooet River, British Columbia (photo by P. Filippelli).

Chapter 16

Pacific Coast Rivers of Canada and Alaska

John S. Richardson and Alexander M. Milner

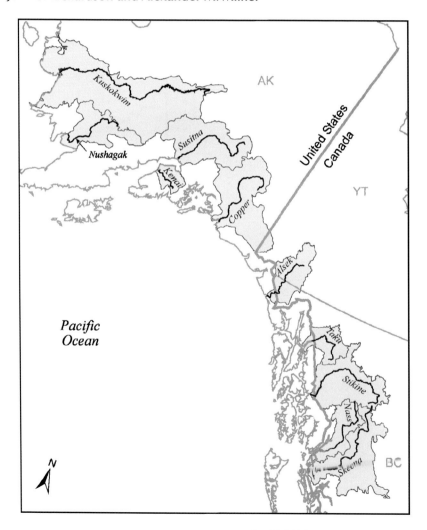

DOI: 10.1016/B978-0-12-375088-4.00016-6

The Pacific Coast rivers of Canada and Alaska (see map) cover the region extending from British Columbia's north coast (north of the Fraser River, Chapter 15) to north of the Aleutian Islands (south of the Yukon River, Chapter 17), and have a latitudinal range from 54°N to 64°N. The southernmost rivers in this region have their catchments entirely within British Columbia, and the northernmost rivers are within Alaska. Several Transboundary Rivers have a large portion of their catchment within Canada before flowing through Alaska to the Pacific Ocean. Most of the coastal rivers are characterized by catchments originating at high elevations within a series of mountain ranges. These classic wild salmon rivers include some of the most pristine drainages in the north temperate region. Each of the basins described here has their upper drainages in mountainous terrain, some originating from glaciers, and discharge to the ocean. Several have their highest points well above 3000 m within the Alaska Range (including Mt. McKinley), the Wrangell Mountains, the St-Elias Mountains, and the Coast Mountains.

Aboriginal peoples have a close association with most of these large rivers, providing important food resources, notably salmon. It seems likely that human settlement dates to the end of the Wisconsinan glacial period (approximately 11,000 years ago) when the Bering land bridge between Asia and Alaska (Beringia) still existed.[1] Rich aquatic resources and abundant wildlife supported subsistence living in these coastal regions. Large winter villages and a dependence on salmon appear to be among the factors leading to the rich set of traditions and culture associated with coastal peoples. There are many native groups in the region, largely differentiated on the basis of language. Europeans reached these coastal areas during the late 1700s and trading posts and fishing ports were established during the early 1800s. It was not until the gold rushes of the mid-1800s that large numbers of European settlers moved inland. Today the smaller communities of this region are well represented by aboriginal peoples with many traditional occupations, such as trapping and fishing.

We will discuss 10 major rivers from this region, several of which are among the 30 largest rivers in North America. The Kuskokwim, Susitna, Stikine, Skeena, Nushagak, and Copper rivers all have mean discharges $\geq 1000 \, m^3/s$. The Alsek, Taku, and Nass rivers are somewhat smaller, but still have discharges $\geq 600 \, m^3/s$. Although the Kenai River is not as large as many others in the region, it is very important economically, supporting significant commercial sports fisheries.

---

1 McGhee, R. 1996. Ancient People of the Arctic. University of British Columbia Press, Vancouver.

Kuskokwim River

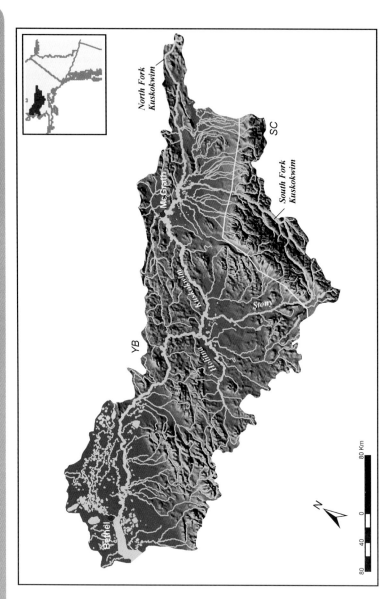

Relief: >3550 m
Basin area: 124,319 km²
Mean discharge: 1900 m³/s
Mean annual precipitation: 42 cm

Mean air temperature: ~1.6°C
Mean water temperature: ~5°C
No. of fish species: 27 to 31
No. of endangered species: 0

Physiographic provinces: Yukon Basin (YB), South Central Alaska (SC)
Major fishes: sockeye salmon, chinook salmon, coho salmon, chum salmon, pink salmon, Pacific lamprey, humpback whitefish, lake trout, Dolly Varden, longfin smelt, boreal smelt, eulachon, longnose sucker, burbot, threespine stickleback, coastrange sculpin
Major other aquatic vertebrates: river otter, muskrat, moose, mink, merganser, belted kingfisher, American dipper
Major benthic insects: NA
Nonnative species: none
Major riparian plants: willows, alder
Special features: globally important wildlife area in Yukon–Kuskokwim delta, especially for breeding

waterfowl; part of Beringian glacial refuge and home to several endemic species; a variety of water sources, including glacier runoff, snowmelt, wetlands, and forest; mostly pristine wilderness
Fragmentation: none

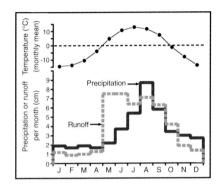

Kuskokwim River, upstream from Bethel, Alaska (photo by L. Renan).

Susitna River

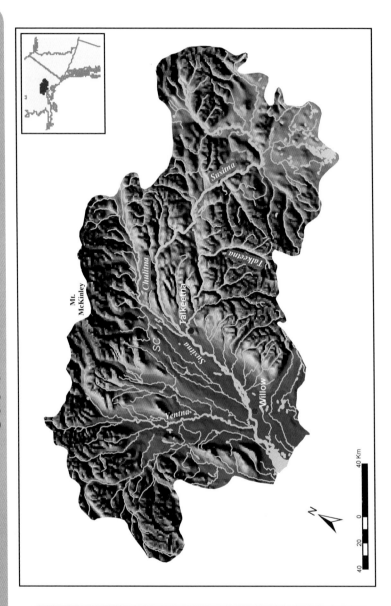

Relief: >4000 m
Basin area: 51,800 km²
Mean discharge: 1427 m³/s
Mean annual precipitation: 71 cm
(underestimate)

Mean air temperature: 0.8°C
Mean water temperature: 4.3°C
No. of fish species: 25 to 29
No. of endangered species: 0

Physiographic province: South Central Alaska (SC)

Major fishes: chinook salmon, coho salmon, chum salmon, pink salmon, sockeye salmon, Bering cisco, round whitefish, rainbow trout, Dolly Varden, Arctic grayling, Arctic lamprey, burbot

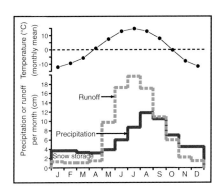

Major other aquatic vertebrates: river otter, muskrat, moose, mink, merganser, belted kingfisher, American dipper

Major benthic insects: mayflies (*Baetis*, *Cinygmula*)

Nonnative species: none

Major riparian plants: willow, alder, cottonwood

Special features: major glacier-fed system that dominates landscape to the southwest of Alaska Range with many complex, off-channel habitats; drains from Mount McKinley, highest point in North America, and other glaciated peaks of the Alaska Range (U.S.)

Fragmentation: none

Susitna River, Alaska, showing different channels (photo by A. Milner).

Kenai River

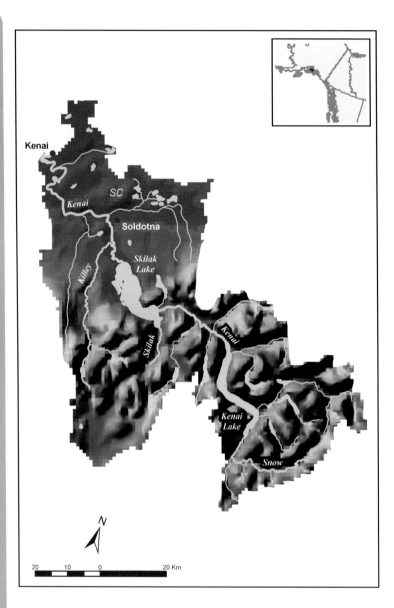

Relief: >1500 m
Basin area: 5206 km^2
Mean discharge: 167.7 m^3/s
Mean annual precipitation: 49 cm
(underestimate)

Mean air temperature: 0.9°C
Mean water temperature: 6.4°C
No. of fish species: 28
No. of endangered species: 0

Physiographic province: South Central Alaska (SC)

Major fishes: Dolly Varden, Arctic grayling, chinook salmon, sockeye salmon, coho salmon, pink salmon, chum salmon, rainbow trout, steelhead, Pacific lamprey, humpback whitefish, lake trout, longfin smelt, boreal smelt, eulachon, longnose sucker, burbot, threespine stickleback, coastrange sculpin

Major other aquatic vertebrates: river otter, muskrat, beaver, moose, mink, merganser, belted kingfisher, American dipper

Major benthic insects: true flies (*Diamesa, Cricotopus/Orthocladius, Pagastia, Parakiefferiella, Eukiefferiella, Rheotanytarsus*), mayflies (*Baetis, Cinygmula, Ephemerella*), stoneflies (*Paraleuctra, Plumiperla, Isoperla*), caddisflies (*Brachycentrus, Ecclisocosmoceus, Glossosoma, Hydropsyche*)

Nonnative species: none

Major riparian plants: willow, alder

Special features: diversity of clearwater, glacier-influenced, and wetland-stained river systems; very productive salmon river; large tourism industry based on freshwater and tidal zone fishing; valuable economies through commercial and sports fisheries

Fragmentation: none

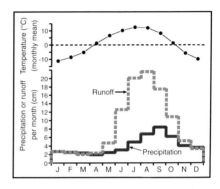

Kenai River, Kenai Peninsula, Alaska (photo by A. Milner).

Stikine River

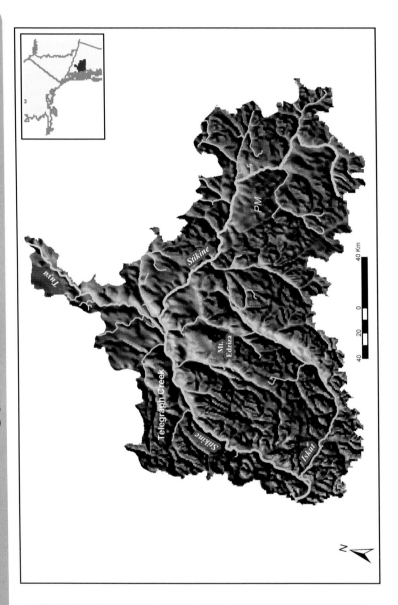

Relief: >2900 m
Basin area: 51,592 km²
Mean discharge: 1587 m³/s
Mean annual precipitation: 38 cm
(underestimate)

Mean air temperature: 2.0°C
Mean water temperature: 6.6°C
No. of fish species: 22 to 26
No. of endangered species: 1

Physiographic province: Coast Mountains of British Columbia and Southeast Alaska (PM)

Major fishes: sockeye salmon, chinook salmon, Arctic grayling, burbot, chum salmon, coho salmon, cutthroat trout, longnose sucker, mountain whitefish, rainbow trout, lake chub, pink salmon, threespine stickleback, sculpins

Major other aquatic vertebrates: river otter, muskrat, beaver, moose, mink, merganser, belted kingfisher, American dipper

Major benthic insects: mayflies (*Baetis, Cinygmula, Rhithrogena, Ameletus*), stoneflies (*Suwallia, Capnia, Doddsia, Isoperla, Podmosta, Zapada*), true flies (*Diamesa, Cricotopus, Orthocladius, Euryhapsis, Eukiefferiella*)

Nonnative species: none; threats of Atlantic salmon; historical records of American shad

Major riparian plants: willows, alders, cottonwood

Special features: one of the largest free-flowing rivers draining a temperate biome; largely pristine wilderness with some glacial drainage; drains Edziza Provincial Park and Spatsizi Plateau Wilderness Provincial Park

Fragmentation: none

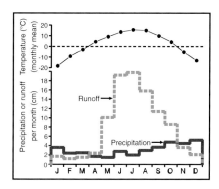

Stikine River, British Columbia (photo by Tim Palmer).

Skeena River

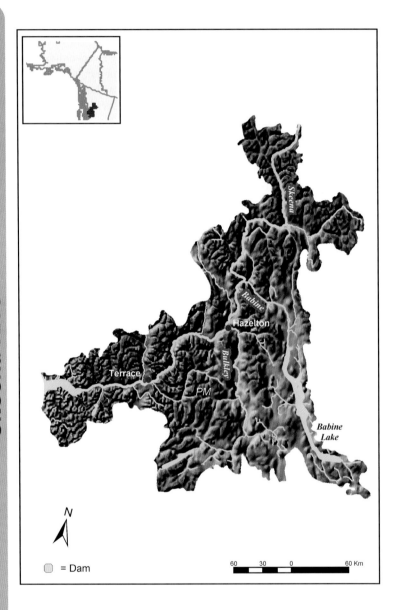

N

⬤ = Dam

60 30 0   60 Km

Relief: 2755 m
Basin area: 54,400 km^2
Mean discharge: 1760 m^3/s
Mean annual precipitation: 63 cm
(underestimate)

Mean air temperature: 4.3°C
Mean water temperature: ~6.0°C
No. of fish species: 33
No. of endangered species: 1

Physiographic province: Coast Mountains of British Columbia and Southeast Alaska (PM)

Major fishes: pink salmon, sockeye salmon, chinook salmon, coho salmon, steelhead, rainbow trout, cutthroat trout, Dolly Varden, eulachon, burbot, mountain whitefish, chum salmon, prickly sculpin, Pacific lamprey, burbot, largescale sucker, peamouth chub, pygmy whitefish, white sucker, white sturgeon, northern pikeminnow

Major other aquatic vertebrates: river otter, muskrat, beaver, moose, mink, merganser, belted kingfisher, American dipper, tailed frog

Major benthic insects: mayflies (*Baetis*, *Rhithrogena*, *Epeorus*, *Cinygmula*, *Drunella*, *Serratella*, *Paraleptophlebia*), stoneflies (*Sweltsa*, *Zapada*), caddisflies (*Rhyacophila*)

Nonnative species: none; threats of Atlantic salmon; historical records of American shad

Major riparian plants: willows, alders, vine maple, western red cedar, western hemlock

Special features: largest free-flowing river in North America draining a temperate biome; productive salmon river

Fragmentation: very low; dam in tributary affects about 0.2% of total flow

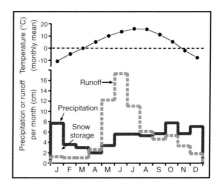

Skeena River below Terrace, British Columbia (photo by Tim Palmer).

Nushagak River

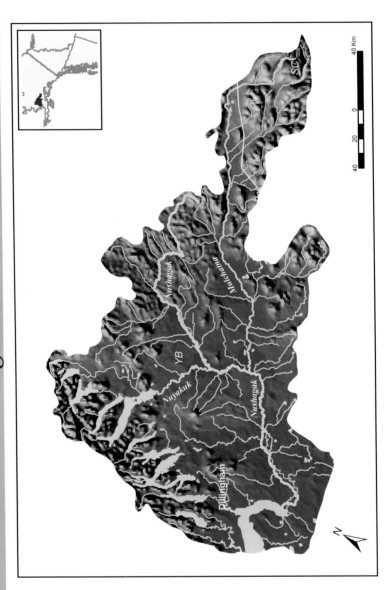

Relief: ~600 m   Mean air temperature: 0.9°C
Basin area: 34,706 km^2 Mean water temperature: NA
Mean discharge: 1000 m^3/s   No. of fish species: 27 to 31
Mean annual precipitation: 66 cm   No. of endangered species: 0
(underestimate)

Physiographic provinces: Yukon Basin (YB), South Central Alaska (SC)
Major fishes: sockeye salmon, chinook salmon, coho salmon, chum salmon, pink salmon, Alaska blackfish, Dolly Varden, northern pike, rainbow trout, slimy sculpin, Arctic grayling, lake trout, threespine stickleback, rainbow smelt, humpback whitefish, Pacific lamprey, burbot
Major other aquatic vertebrates: river otter, muskrat, moose, mink, merganser, belted kingfisher, American dipper
Major benthic insects: mayflies (Ephemerellidae), stoneflies (Chloroperlidae, Nemouridae), caddisflies (Hydroptilidae, Glossosomatidae), true flies (chironomid midges)
Nonnative species: none
Major riparian plants: grasses, willow, alder

Special features: pristine wilderness; Mulchatna and Chilikadrotna tributaries are U.S. National Wild and Scenic Rivers; large glacial lakes are predominant feature of basin; long-term studies of salmon and their food webs by University of Washington at Woods Lakes research station
Fragmentation: none

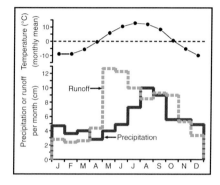

Nushagak River near its confluence with Mulchatna River, Alaska (photo by E. McKittrick, Ground Truth Trekking).

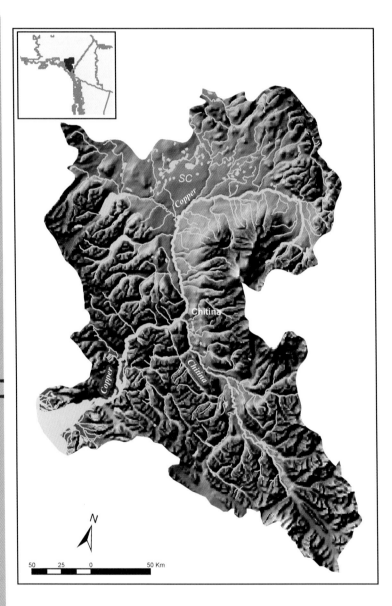

Copper River

Relief: >2500 m
Basin area: 63,196 km²
Mean discharge: 1785 m³/s
Mean annual precipitation: 28 cm
(underestimate)

Mean air temperature: −2.7°C
Mean water temperature: NA
No. of fish species: 24 to 27
No. of endangered species: 0

Physiographic province: South Central Alaska (SC)

Major fishes: coho salmon, longnose sucker, arctic grayling, Dolly Varden, chinook salmon, sockeye salmon, chum salmon, round whitefish, slimy sculpin

Major other aquatic vertebrates: river otter, muskrat, beaver, moose, mink, merganser, belted kingfisher, American dipper

Major benthic insects: true flies (*Diamesa*, black flies), mayflies (Baetidae, Ameletidae, Ephemerellidae, Heptageniidae), stoneflies (Nemouridae, Chloroperlidae), caddisflies (Lepidostomatidae)

Nonnative species: none

Major riparian plants: willows, alders

Special features: Gulkana River, a tributary, is a U.S. National Wild and Scenic River; drains from Wrangell–St. Elias National Park (U.S.) and Kluane National Park (Canada)

Fragmentation: none

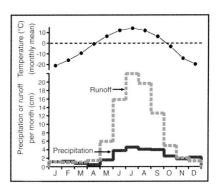

Copper River (left to right) at confluence with Chitina River (center in distance), Chitina, Alaska (photo by M. C. T. Smith).

Alsek River

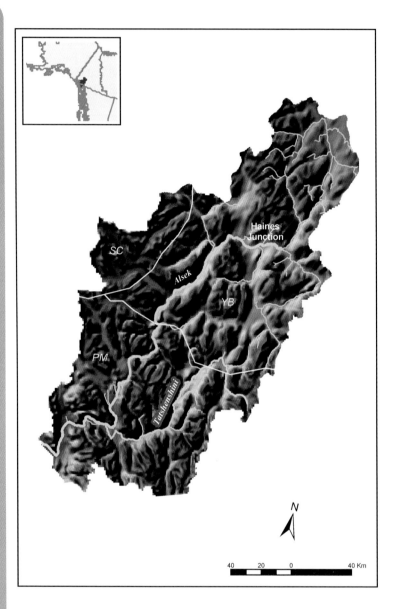

Relief: >3000 m
Basin area: 28,023 km²
Mean discharge: 862.6 m³/s
Mean annual precipitation: 31 cm
(underestimate)

Mean air temperature: −2.9°C
Mean water temperature: NA
No. of fish species: ≥32
No. of endangered species: 0

Physiographic provinces: South Central Alaska (SC), Coast Mountains of British Columbia and Southeast Alaska (PM), Yukon Basin (YB)

Major fishes: Arctic grayling, chinook salmon, coho salmon, cutthroat trout, Dolly Varden, sockeye salmon, steelhead, chum salmon, burbot, northern pike, lake trout

Major other aquatic vertebrates: river otter, muskrat, beaver, moose, mink, merganser, belted kingfisher, American dipper

Major benthic insects: true flies (*Diamesa, Cricotopus, Rheotanytarsus, Eukiefferiella, Thienemannimyia*), mayflies (*Cinygmula, Baetis, Ephemerella, Rhithrogena, Drunella*), stoneflies (*Sweltsa, Capnia*)

Nonnative species: none; threats of Atlantic salmon; historical records of American shad

Major riparian plants: willows, alders

Special features: most of basin protected within wilderness areas and National Parks; Alsek and its major tributary, the Tatshenshini River, are part of Canadian Heritage River system and part of World Heritage Site (UNESCO); popular for river rafting

Fragmentation: one tributary has dam affecting ~2% of Alsek flow in live storage

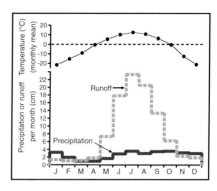

Alsek River, British Columbia (photo by Tim Palmer).

Taku River

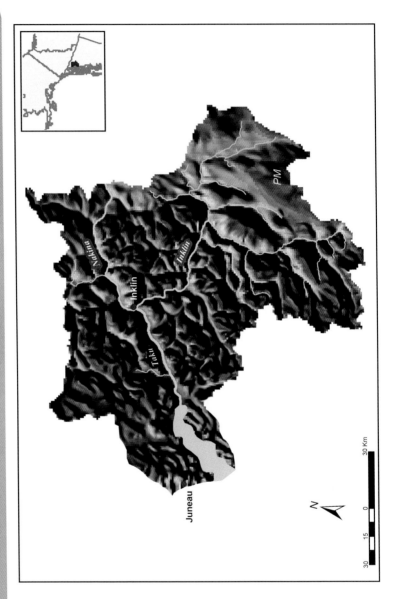

Relief: >2300 m   Mean air temperature: 0°C
Basin area: 29,800 km²   Mean water temperature: NA
Mean discharge: 600 m³/s   No. of fish species: ≥32
Mean annual precipitation: 34 cm   No. of endangered species: 0
(underestimate)

Physiographic province: Coast Mountains of British Columbia and Southeast Alaska (PM)

Major fishes: sockeye salmon, pink salmon, coho salmon, chinook salmon, Dolly Varden, cutthroat trout, steelhead, chum salmon, eulachon, longfin smelt, Pacific lamprey, round whitefish, slimy sculpin, threespine stickleback

Major other aquatic vertebrates: river otter, muskrat, beaver, moose, mink, merganser, belted kingfisher, American dipper

Major benthic insects: NA

Nonnative species: none; threats of Atlantic salmon; historical records of American shad

Major riparian plants: alders, willows

Special features: primarily wilderness basin; peak flows fill Tulsequah Lake and cause lake outburst events most years (as lake swells the water can float the glacier or melt it sufficiently to cause sudden releases); popular river for rafting; very productive salmon river

Fragmentation: none

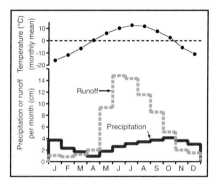

Taku River between Juneau and Taku Inlet (photo by C. Holland).

Nass River

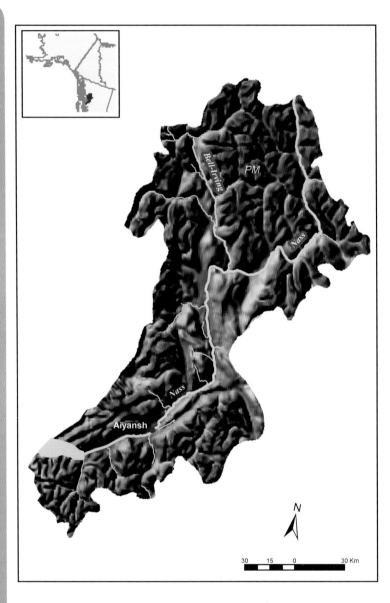

Relief: 2368 m
Basin area: 21,100 km²
Mean discharge: 892 m³/s
Mean annual precipitation: 130 cm

Mean air temperature: 6.1°C
Mean water temperature: NA
No. of fish species: 27
No. of endangered species: 0

Physiographic province: Coast Mountains of British Columbia and Southeast Alaska (PM)

Major fishes: chinook salmon, steelhead, chum salmon, coho salmon, sockeye salmon, pink salmon, cutthroat trout, eulachon, mountain whitefish, Dolly Varden, Pacific lamprey, threespine stickleback

Major other aquatic vertebrates: river otter, muskrat, beaver, moose, mink, merganser, belted kingfisher, American dipper

Major benthic insects: NA

Nonnative species: none; threats of Atlantic salmon; historical records of American shad

Major riparian plants: willows, alders, cottonwood

Special features: largely pristine catchment draining extensive areas in lee of Coast Range Mountains; flows 380km from most inland source to Pacific; drains large areas of high-elevation forest and tundra; productive salmon river

Fragmentation: none

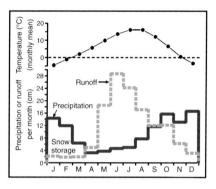

Nass River from Gitwinksihikw Bridge (photo by K. Freeman).

Yukon River Basin

Robert C. Bailey

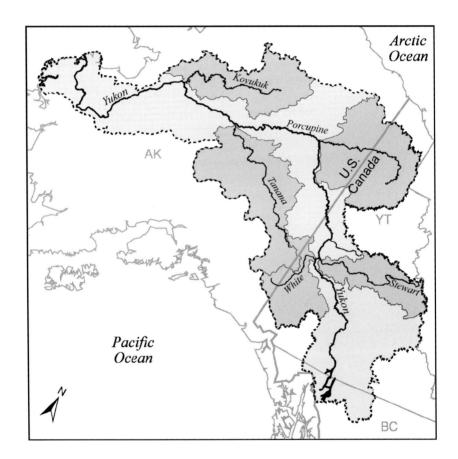

DOI: 10.1016/B978-0-12-375088-4.00017-8

The Yukon basin is the 7th largest in North America (after the Mississippi, Missouri, St. Lawrence, Nelson, Mackenzie, and Rio Grande), with an area of 839,200 km². Approximately 61% of the basin is in Alaska and 39% in Canada (see map). Most (>90%) of the Canadian portion of the basin is in the Yukon Territory, although many of the headwater lakes and streams are in northern British Columbia. The Yukon River is one of the wildest and is by far the longest free-flowing river in North America, with only a single dam at its headwaters and one each on two of its major tributaries. From its headwater lakes less than 30 km from the Pacific Ocean in British Columbia, to its outflow into the Bering Sea 3200 km downstream in Alaska, the vast Yukon River and its basin encompass a heterogeneous collection of often spectacular ecosystems, from ephemeral mountain streams to mini deserts, from opaque, glacial runoff rivers to clear, boreal forest creeks, from torrential running waters to productive wetland "flats" of 1000s of km² of small lakes and ponds. The southern origin of the Yukon River, draining the eastern side of the Coast Mountains and the west side of the Rockies in northern British Columbia, is at a latitude of 59°N and a longitude (133°W) that is about 15° west of Los Angeles. The Bering Sea outflow of the river at 63°N, is 30° west (164°W) of the headwaters, and the river basin itself extends to a northern limit of about 68°N in the Porcupine River watershed.

Although sparsely populated, the Yukon River Basin has been both prehistorically and more recently very important to the human population of North America and the world. Most anthropologists agree that the first human colonization of North America was via a land bridge that crossed what is now Bering Strait >13,000 years ago.[1] Subsequent to the colonization, the basin became an important, resource rich environment for the development and evolution of northern aboriginal communities. In the last 300 years, the Yukon River Basin has attracted fur traders and gold miners from Russia, southern North America, and around the world. More recently, and not without controversy, the basin has been part of a transportation route to the south for oil and gas rather than humans.

In this chapter, I describe the main-stem Yukon as well as 5 of its major tributaries, the Tanana, Koyukuk, White, Stewart, and Porcupine rivers, all with discharges over 400 m³/s (see map).

---

1 Yesner, D. R. 2001. Human dispersal into interior Alaska: antecedent conditions, mode of colonization, and adaptations. Quaternary Science Reviews 20: 315–327.

Yukon River

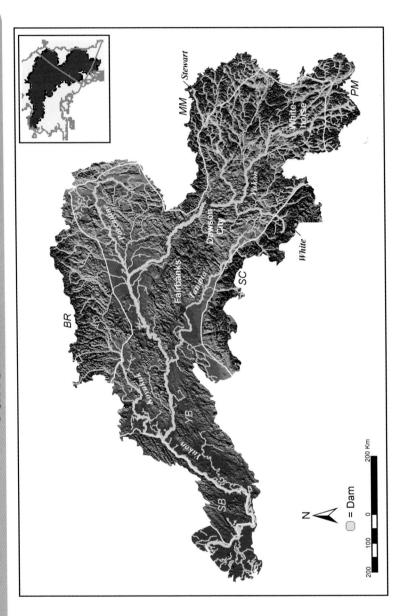

Relief: 6200 m
Basin area: 839,200 km²
Mean discharge: 6340 m³/s
Mean annual precipitation: 33 cm

Mean air temperature: −3°C
Mean water temperature: 7.7°C
No. of fish species: 30
No. of endangered species: 0

Physiographic provinces: Yukon Basin (YB), Seward Peninsula and Bering Coast Uplands (SB), Brooks Range (BR), Mackenzie Mountains (MM), South-Central Alaska (SC), Coast Mountains of British Columbia and Alaska (PM)

Major fishes: Bering cisco, whitefish, chinook salmon, chum salmon, coho salmon, Arctic grayling, inconnu (sheefish), lake trout, Alaska blackfish, northern pike, burbot, Dolly Varden, Arctic lamprey

Major other aquatic vertebrates: muskrat, brown bear, black bear, mink, river otter, bald eagle

Major benthic insects: mayflies (*Baetis, Epeorus, Heptagenia, Siphlonurus, Drunella, Leptophlebia*), stoneflies (*Alaskaperla, Capnia, Eucapnopsis, Utacapnia, Zapada, Taenionema, Alloperla, Suwallia,*

and Isoperla), true flies (chironomid midges, *Simulium, Gymnopais*)

Nonnative species: rainbow trout, threespine stickleback

Major riparian plants: mountain alder, water birch, black spruce

Special features: largest wild river in North America

Fragmentation: free-flowing except for dam at Whitehorse

Upper Yukon River at Rink Rapids between Whitehorse and Dawson City, Yukon Territory (photo by Tim Palmer).

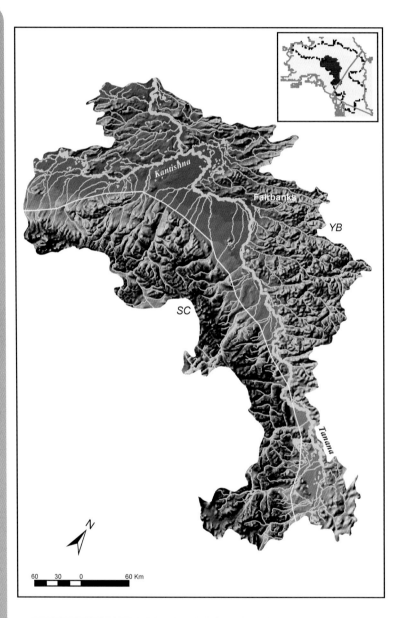

Relief: 6100 m
Basin area: 114,000 km²
Mean discharge: 1185 m³/s
Mean annual precipitation: 28 cm

Mean air temperature: −2.9°C
Mean water temperature: 6,7°C
No. of fish species: 20
No. of endangered species: 0

Physiographic provinces: South-Central Alaska (SC), Yukon Basin (YB)

Major fishes: Bering cisco, whitefish, chinook salmon, chum salmon, coho salmon, Arctic grayling, inconnu (sheefish), lake trout, Alaska black-fish, northern pike, burbot

Major other aquatic vertebrates: brown bear, black bear, beaver, mink, marten, river otter, bald eagle

Major benthic insects: unconfirmed; see Yukon River

Nonnative species: rainbow trout, Arctic char

Major riparian plants: black spruce, mountain alder willow, quaking aspen

Special features: mixture of turbid glacial input from the south and clear subsurface inflow from the north

Fragmentation: free-flowing except for flood-control dam on Chena River

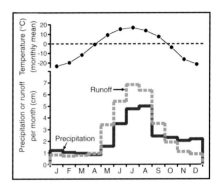

Tanana River above Tok, Alaska (photo by Tim Palmer).

Koyukuk River

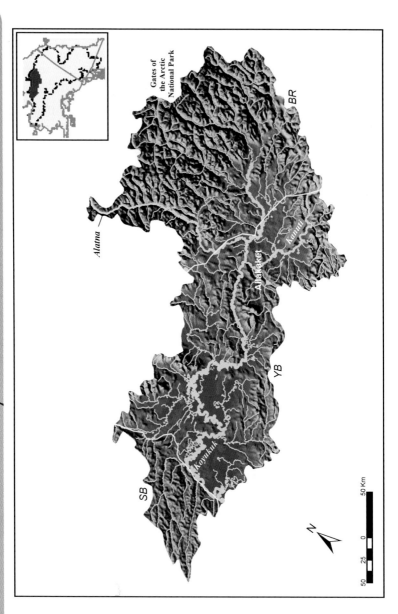

Relief: 2280 m
Basin area: 83,500 km^2
Mean discharge: 664 m^3/s
Mean annual precipitation: 31 cm

Mean air temperature: −9.5°C
Mean water temperature: 8.6°C
No. of fish species: ~25
No. of endangered species: 0

Physiographic provinces: Brooks Range (BR), Yukon Basin (YB), Seward Peninsula and Bering Coast Uplands (SB)

Major fishes: unconfirmed; see Yukon River

Major other aquatic vertebrates: see Yukon River

Major benthic insects: unconfirmed; see Yukon River

Nonnative species: rainbow trout

Major riparian plants: black spruce, mountain alder, balsam poplar, quaking aspen, willow

Special features: almost pristine tundra river; important spawning areas for anadromous fishes; headwaters contact Trans-Alaskan pipeline

Fragmentation: none

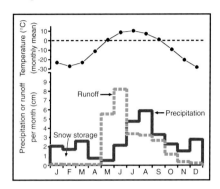

South Fork of the Koyukuk River, from the Dalton Highway, Alaska (photo by D. Shiozawa).

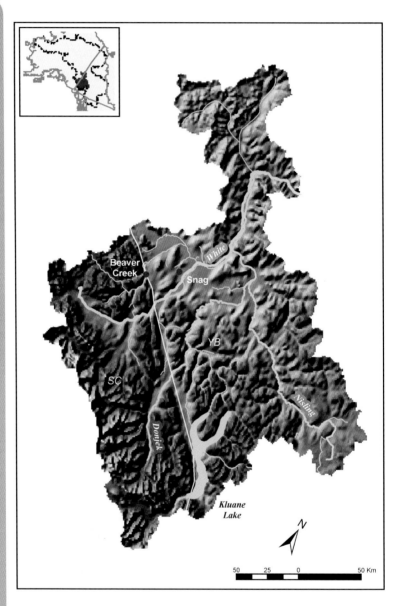

White River

Relief: 5635 m
Basin area: 50,500 km²
Mean discharge: 927 m³/s
Mean annual precipitation: 31 cm

Mean air temperature: 3.8°C
Mean water temperature: NA
No. of fish species: 28
No. of endangered species: 0

Physiographic provinces: South-Central Alaska (SC), Yukon Basin (YB)

Major fishes: Bering cisco, whitefish, chinook salmon, chum salmon, coho salmon, Arctic grayling, inconnu (sheefish), lake trout, Alaska blackfish, northern pike, burbot

Major other aquatic vertebrates: see Yukon River

Major benthic insects: unconfirmed; see Yukon River

Nonnative species: none

Major riparian plants: black spruce, mountain alder, quaking aspen, balsam poplar

Special features: pristine river with primarily glacial inputs; extremely turbid

Fragmentation: none

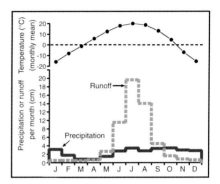

White River at the Alaska Highway near Koidern, Yukon Territory (photo by R. Bailey).

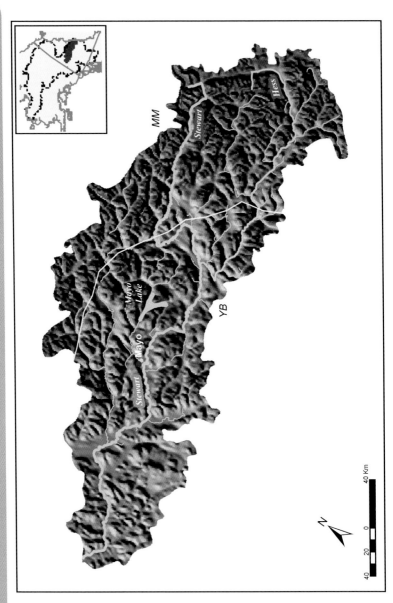

Stewart River

Relief: 2635 m
Basin area: 51,000 km²
Mean discharge: 675 m³/s
Mean annual precipitation: 28 cm

Mean air temperature: − 2.9°C
Mean water temperature: NA
No. of fish species: ~29
No. of endangered species: 0

Physiographic provinces: Mackenzie Mountains (MM), Yukon Basin (YB)

Major fishes: Bering cisco, whitefish, chinook salmon, chum salmon, coho salmon, Arctic grayling, inconnu (sheefish), lake trout, northern pike, burbot

Major other aquatic vertebrates: see Yukon River

Major benthic insects: unconfirmed; see Yukon River

Nonnative species: rainbow trout, lake trout

Major riparian plants: see Yukon River

Special features: mostly pristine river, heavy placer activity in lower Stewart River tributaries

Fragmentation: none; one dam on tributary (Mayo River)

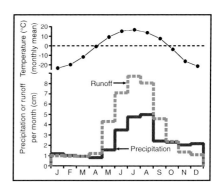

Stewart River looking southwest from Mayo, Yukon Territory (photo by H. Burian).

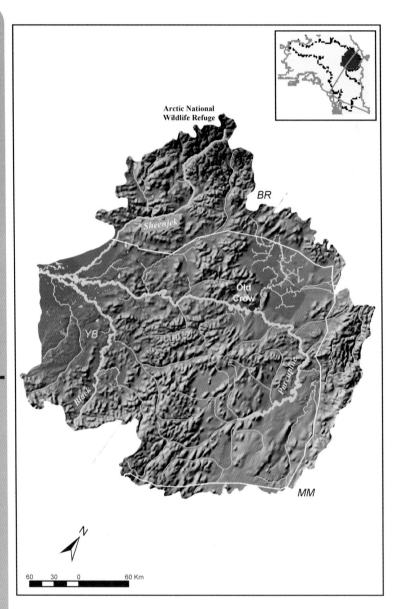

Porcupine River

Relief: 1670 m
Basin area: 118,000 km²
Mean discharge: 414 m³/s
Mean annual precipitation: 16 cm

Mean air temperature: −11°C
Mean water temperature: 5.5°C
No. of fish species: ~29
No. of endangered species: 0

Physiographic provinces: Mackenzie Mountains (MM), Brooks Range (BR), Yukon Basin (YB)

Major fishes: Bering cisco, whitefish, chinook salmon, chum salmon, coho salmon, Arctic grayling, inconnu (sheefish), lake trout, Alaska blackfish, northern pike, burbot

Major other aquatic vertebrates: brown bear, black bear, beaver, mink, marten, river otter, bald eagle

Major benthic insects: unconfirmed; see Yukon River

Nonnative species: rainbow trout, threespine stickleback

Major riparian plants: black spruce, mountain alder

Special features: large pristine tundra river; northernmost basin within Yukon system

Fragmentation: none

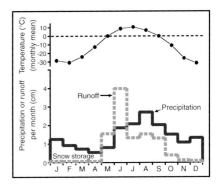

Porcupine River, Yukon Flats National Wildlife Refuge, Alaska (photo by D. Spencer).

Chapter 18

Mackenzie River Basin

Joseph M. Culp, Terry D. Prowse, and Eric A. Luiker

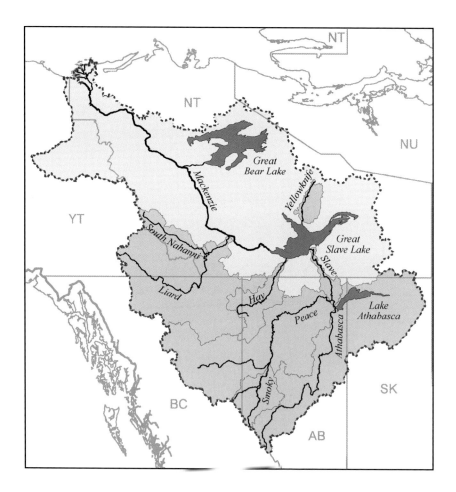

The Mackenzie River Basin encompasses an enormous geographic area extending over 15° of latitude and 37° of longitude from central Alberta to the Arctic Ocean, an area (1,787,000 km^2) larger than central Europe (see map). The Mackenzie is the second largest river basin in North America and its discharge (9020 m^3) ranks third, following only the Mississippi and St. Lawrence rivers. The system includes a number of major rivers as well as the main-stem channel that empties through the Mackenzie Delta, the second-largest Arctic delta in the world, into the Beaufort Sea. In addition, the system contains two large freshwater deltas (the Peace-Athabasca and the Slave) and three major lakes (Lake Athabasca, Great Slave Lake, and Great Bear Lake).

The first inhabitants in the Mackenzie watershed area were the ancestral American Indians, who moved to the area from Asia across the Bering Sea at least 12,000 years ago. Immigration by the Inuit followed and they eventually occupied arctic coastal areas including the Mackenzie Delta and other tundra areas north of the tree line. Early contacts between the Aboriginal Peoples of the Mackenzie River Basin and Europeans appear to have occurred in the late 1600s and early 1700s as a result of fur trading. Many of the early explorers worked for the North West Company or the Hudson's Bay Company, the two businesses being amalgamated in 1821. Several of these explorers, including David Thompson and Simon Fraser, travelled through parts of the basin in search of new trade routes to the Pacific and the North. However, the river bears the name of Sir Alexander Mackenzie because he explored the mainstem from Great Slave Lake to its Arctic outlet in 1789. The Mackenzie River main stem from Great Slave Lake to the Beaufort Sea has been a key transportation route for the settlement of the region with wood burning paddle wheelers operating on the river from the 1880s until the 1940s. Barge traffic remains an important method of transporting commodities and supplies to northern communities.

In this chapter, we discuss the main-stem Mackenzie as well as nine of its tributaries that illustrate the large range of natural diversity and human impacts within the basin. The largest rivers include the Athabasca and Peace that join to form the Slave (3437 m^3/s), all of which drain the most southern reaches. The other extremely large river is the Liard (2446 m^3/s) that joins the Mackenzie main stem to the north. Other rivers include the relatively large (>300 m^3/s) Smoky and South Nahanni rivers, and the smaller Hay and Yellowknife rivers.

Mackenzie River

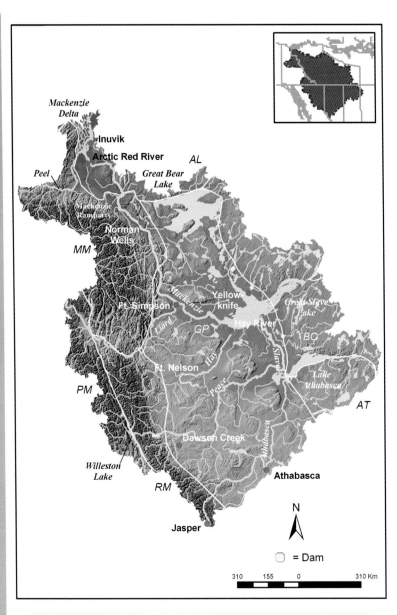

Relief: 3620 m
Basin area: 1,787,000 km²
Mean discharge: 9020 m³/s
Mean annual precipitation: 26 cm

Mean air temperature: −9.5°C
Mean water temperature; 5.3°C
No. of fish species: 52 (entire basin)
No. of endangered species: 0

Physiographic provinces: Mackenzie Mountains (MM), Coast Mountains of British Columbia and Southeast Alaska (PM), Rocky Mountains in Canada (RM), Great Plains (GP), Athabasca Plains (AT), Bear–Slave–Churchill Uplands (BC), Arctic Lowlands (AL)
Major fishes: Arctic lamprey, goldeye, Arctic cisco, lake cisco, Arctic char, least cisco, lake whitefish, broad whitefish, mountain whitefish, pond smelt, rainbow smelt, lake chub, flathead chub, longnose dace, inconnu, Arctic grayling, lake trout, northern pike, longnose suckers, white sucker, troutperch, burbot, slimy sculpin, spoonhead sculpin, walleye
Major other aquatic vertebrates: wood frog, beluga whale (Mackenzie Delta), muskrat, moose, mink, beaver, river otter, snow goose, black brant, greater white-fronted goose, tundra swan
Major benthic insects: true flies (Ceratopogonidae, Chironomidae, Simuliidae), mayflies (*Ametropus*, *Baetis*, *Ephemerella*, *Heptagenia*), stoneflies (*Isoperla*), caddisflies (*Brachycentrus*)
Nonnative species: none
Major riparian plants: horsetail, bulrush, cattail, Labrador tea, willow, sedges, balsam poplar, white spruce, black spruce, alder, birch, tamarack
Special features: largest river in Canada; Limestone Canyon of the Ramparts, Mackenzie Delta, Great Bear Lake, Great Slave Lake
Fragmentation: none on main stem

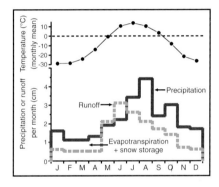

Mackenzie River (left) at confluence with Liard River (right) (photo by T. D. Prowse).

Liard River

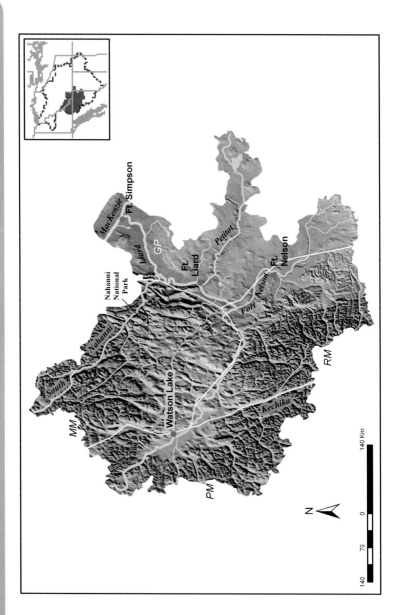

Relief: 2573 m
Basin area: 277,000 km²
Mean discharge: 2446 m³/s
Mean annual precipitation: 36 cm

Mean air temperature: −3.7°C
Mean water temperature: 6.1°C
No. of fish species: 34
No. of endangered species: 1

Physiographic provinces: Coast Mountains of British Columbia and Southeast Alaska (PM), Rocky Mountains in Canada (RM), Great Plains (GP), Mackenzie Mountains (MM)

Major fishes: chum salmon, Arctic cisco, mountain whitefish, lake whitefish, inconnu, bull trout, longnose sucker, white sucker, Arctic grayling, goldeye, northern pike, walleye, burbot, lake chub, flathead chub, northern squawfish, longnose dace, troutperch, slimy sculpin, spoonhead sculpin

Major other aquatic vertebrates: wood frog, moose, beaver, river otter, muskrat, mink, trumpeter swan (nesting population at Yohin Lake), common loon, red-necked grebe, common goldeneye, bald eagle

Major benthic insects: true flies (Diamesinae, Empididae, Orthocladiinae, Simuliidae, Tanyderidae, Tipulidae), mayflies (Baetidae, Ephemerellidae, Heptageniidae), stoneflies (Capniidae, Chloroperlidae, Nemouridae, Perlidae), caddisflies (Brachycentridae, Limnephilidae, Rhyacophilidae)

Nonnative species: none

Major riparian plants: black spruce, white spruce, trembling aspen, balsam poplar, aspen, lodgepole pine, willow, sedges

Special features: Grand Canyon of the Liard, Liard Hot Springs, Hell Gate Rapids, Virginia Falls, Nahanni National Park

Fragmentation: none

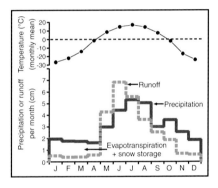

Liard River, Northwest Territories, illustrating ice-jam flooding (photo by T. D. Prowse).

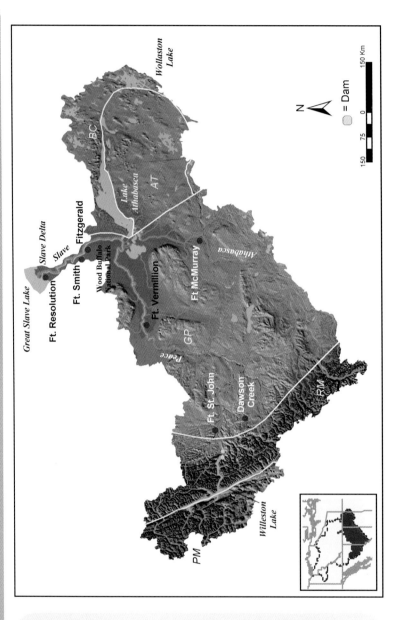

Slave River

Relief: 3500 m
Basin area: 615,000 km^2
Mean discharge: 3437 m^3/s
Mean annual precipitation: 35 cm

Mean air temperature: −3.0°C
Mean water temperature: NA
No. of fish species: 28 (main
stem), 45 (basin)
No. of endangered species: 0

Physiographic provinces: Coast
Mountains of British Columbia
and Southeast Alaska (PM), Rocky
Mountains in Canada (RM), Great
Plains (GP), Athabasca Plain (AT),
Bear–Slave–Churchill Uplands (BC)
Major fishes: Arctic lamprey, goldeye,
northern pike, Dolly Varden, lake
cisco, lake trout, longnose sucker,
white sucker, troutperch, lake
whitefish, walleye, inconnu, flathead
chub, spottail shiner, pearl dace,
burbot, ninespine stickleback, slimy
sculpin, yellow perch
Major other aquatic vertebrates:
chorus frog, wood frog, northern
leopard frog, beaver, muskrat, river
otter, mink, whooping crane, white
pelican, snow goose, Canada goose,
common loon, red-necked grebe,
common merganser, osprey, bald
eagle
Major benthic insects: true flies
(*Stictochironomus, Procladius,
Chironomus, Polypedilum*)
Nonnative species: NA

Major riparian plants: white spruce,
black spruce, tamarack, balsam poplar,
aspen, Labrador tea, reindeer lichens,
peat mosses, sago weed, pickerelweed,
river horsetail, sedges, cattails, reed
grasses, rushes, burrweed, willow, alder
Special features: Wood Buffalo
National Park (whooping crane
breeding sites), Slave River Delta,
Peace–Athabasca Delta
Fragmentation: submerged weir on
the Riviere des Rochers at Little
Rapids and Revillion Coupe; major
dam on (Peace River)

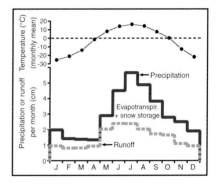

Slave River (photo by M. Conly).

Peace River

Relief: 2130 m
Basin area: 293,000 km²
Mean discharge: 2118 m³/s
Mean annual precipitation: 31 cm

Mean air temperature: −0.9°C
Mean water temperature: NA
No. of fish species: 31
No. of endangered species: 0

Physiographic provinces: Great Plains (GP), Rocky Mountains in Canada (RM), Coast Mountains of British Columbia and Southeast Alaska (PM)

Major fishes: goldeye, walleye, burbot, northern pike, rainbow trout, Dolly Varden, lake trout, mountain whitefish, lake whitefish, lake chub, flathead chub, northern squawfish, longnose dace, trout perch, prickly sculpin, spoonhead sculpin

Major other aquatic vertebrates: long toed salamander, spotted frog, wood frog, beaver, muskrat, river otter, mink, snow goose, Canada goose, trumpeter swan, common loon, red-necked grebe, common merganser, osprey, bald eagle

Major benthic insects: true flies (Orthocladiinae, Tanypodinae, Tanytarsini), stoneflies (*Isoperla*, *Isogenoides*, Capniidae, Taeniopterygidae), mayflies (*Baetis*, *Ephemerella*, *Heptagenia*, *Isonychia*, *Rhithrogena*)

Nonnative species: spottail shiner, fathead minnow, westslope cutthroat trout, brook trout

Major riparian plants: black spruce, trembling aspen, balsam poplar, willow, sedges, wheat grass, sedge, horsetail, Labrador tea

Special features: Vermilion Chutes, Boyer Rapids, Peace–Athabasca Delta

Fragmentation: W.A.C. Bennett Dam at Williston Lake on main stem

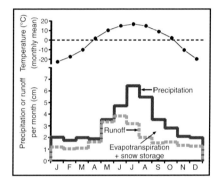

Peace River at Fort Vermilion.

Athabasca River

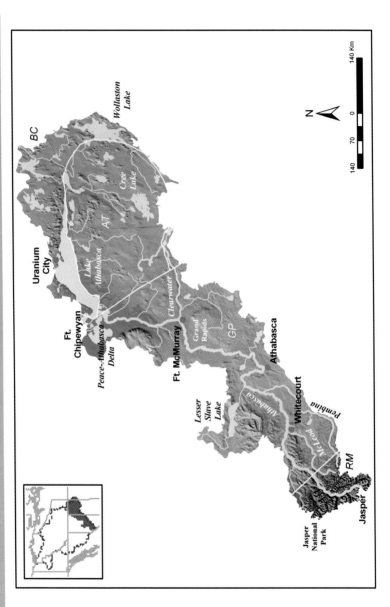

Relief: 3420 m
Basin area: 154,880 km²
Mean discharge: 783 m³/s
Mean annual precipitation: 47 cm

Mean air temperature: 0.2°C
Mean water temperature: 8.2°C
No. of fish species: 36
No. of endangered species: 0

Physiographic provinces: Athabasca Plain (AT), Bear–Slave–Churchill Uplands (BC), Great Plains (GP), Rocky Mountains in Canada (RM)
Major fishes: lake cisco, lake whitefish, round whitefish, mountain whitefish, rainbow trout, Dolly Varden, walleye, lake trout, goldeye, northern pike, burbot, Arctic grayling, longnose sucker, white sucker, flathead chub, longnose dace, troutperch, slimy sculpin, brook stickleback, lake chub
Major other aquatic vertebrates: boreal chorus frog, western toad, wood frog, northern leopard frog, long toed salamander, mink, river otter, moose, muskrat, beaver, greater yellowlegs (sandpiper), killdeer, sandhill crane, bald eagle, osprey, American white pelican, double-crested cormorant, red-necked grebe, great blue heron, common loon, Canada goose, tundra swan
Major benthic insects: mayflies (*Ameletus, Baetis, Cinygmula, Heptagenia, Ephemerella, Rhithrogena*), stoneflies (*Alloperla, Capnia,*

Isoperla, Hesperoperla, Pteronarcella) caddisflies (*Brachycentrus, Glossosoma, Hydropsyche*)
Nonnative species: brown trout, brook trout
Major riparian plants: white spruce, balsam poplar, tamarack, black spruce, aspen willow, water sedge, marsh reed grass, Labrador tea, strawberry-blite
Special features: Athabasca Falls, Grand Rapids, Peace–Athabasca Delta, Canadian Heritage River
Fragmentation: none

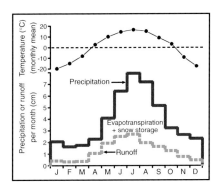

Athabasca River in Jasper National Park, Alberta (photo by C. E. Cushing).

South Nahanni River

Relief: 2573 m
Basin area: 33,388 km^2
Mean discharge: 404 m^3/s
Mean annual precipitation: 36 cm

Mean air temperature: −3.7°C
Mean water temperature: NA
No. of fish species: 13
No. of endangered species: 0

Physiographic provinces: Mackenzie Mountains (MM), Great Plains (GP)

Major fishes: Arctic grayling, Dolly Varden, lake trout, northern pike, lake whitefish, longnose sucker, round whitefish, inconnu

Major other aquatic vertebrates: wood frog, muskrat, beaver, mink, river otter, moose, trumpeter swan, bald eagle, golden eagle, common loon, red-necked grebe

Major benthic insects: NA

Nonnative species: NA

Major riparian plants: white spruce, poplar, black spruce, sedges, wild mint, goldenrod, yellow monkey flower

Special features: Virginia Falls, Rabbitkettle Hotsprings, Sandblowouts, Nahanni National Park, Canadian Heritage River

Fragmentation: none

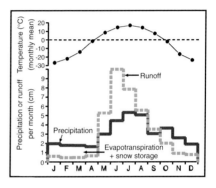

South Nahanni River at Last Chance Harbor above Virginia Falls (photo by D. Bicknell).

Smoky River

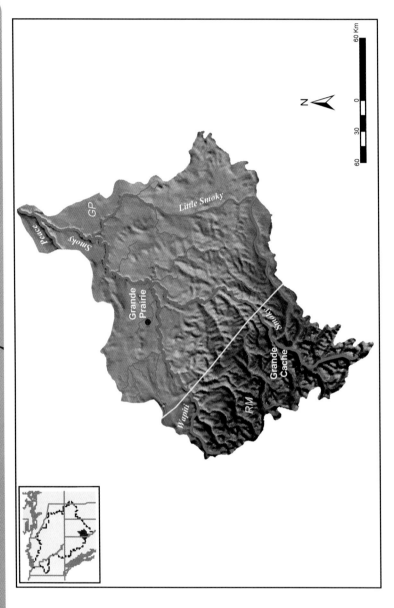

Relief: 2605 m
Basin area: 49,584 km²
Mean discharge: 347 m³/s
Mean annual precipitation: 45 cm

Mean air temperature: 1.6°C
Mean water temperature: NA
No. of fish species: 14
No. of endangered species: 0

Physiographic provinces: Rocky Mountains in Canada (RM), Great Plains (GP)

Major fishes: longnose sucker, white sucker, walleye, northern pike, mountain whitefish, Arctic grayling, Dolly Varden, lake chub, longnose dace, redside shiner, pearl dace, burbot, slimy sculpin, spoonhead sculpin

Major other aquatic vertebrates: wood frog, moose, beaver, muskrat, mink, river otter, American white pelican, great blue heron, trumpeter swan, tundra swan, snow goose, common loon, red-necked grebe, common goldeneye

Major benthic insects: true flies (Chironomini, Orthocladiinae, Tanypodinae), mayflies (*Baetis, Cinygmula, Drunella, Rhithrogena*), stoneflies (*Alloperla, Capnia, Isoperla, Hesperoperla, Pteronarcella, Taenionema*), caddisflies (*Brachycentrus, Glossosoma, Hydropsyche*)

Nonnative species: NA

Major riparian plants: trembling aspen, balsam poplar, black spruce, willow, sedges, wheat grass, sedge, Labrador tea, peat moss, brown moss

Special features: Willmore Wilderness Park, dinosaur tracks (Grande Cache area)

Fragmentation: none

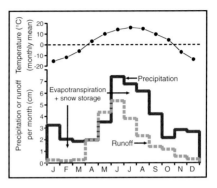

Smoky River, east of Grande Prairie, Alberta.

Hay River

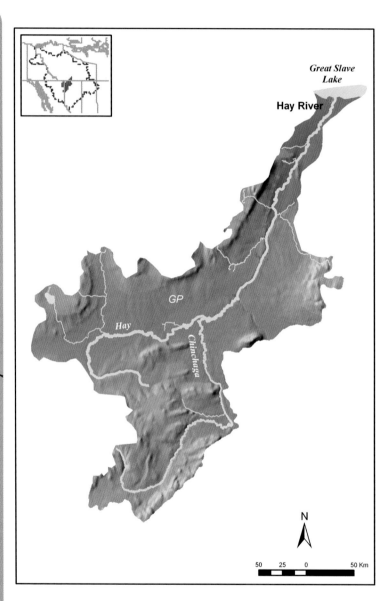

Relief: 1060 m
Basin area: 47,900 km²
Mean discharge: 113 m³/s
Mean annual precipitation: 34 cm

Mean air temperature: −3.4°C
Mean water temperature: 6.4°C
No. of fish species: ~18
No. of endangered species: 0

Physiographic province: Great Plains (GP)

Major fishes: lake chub, flathead chub, longnose dace, longnose sucker, white sucker, northern pike, lake whitefish, mountain whitefish, Arctic grayling, Arctic char, lake trout, walleye, burbot, brook stickleback, slimy sculpin, spoonhead sculpin

Major other aquatic vertebrates: wood frog, moose, beaver, river otter, muskrat, mink, snow goose, Canada goose, osprey, bald eagle, Arctic tern, herring gull, northern pintail

Major benthic insects: NA

Nonnative species: NA

Major riparian plants: balsam poplar, black spruce, white spruce, tamarack, dwarf birch, willows, sedges, Labrador tea, peat moss, brown moss

Special features: Zama and Hay Lakes Wetland area (IBA), Hay River Gorge, Twin Falls; Hay River (community) is northernmost railroad accessible area in Northwest

Territories and important hub for barge travel along Mackenzie River

Fragmentation: none

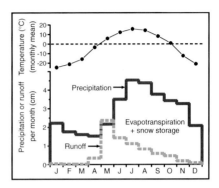

Hay River, south of the town of Hay River, Northwest Territories (photo by T. Carter).

Yellowknife River

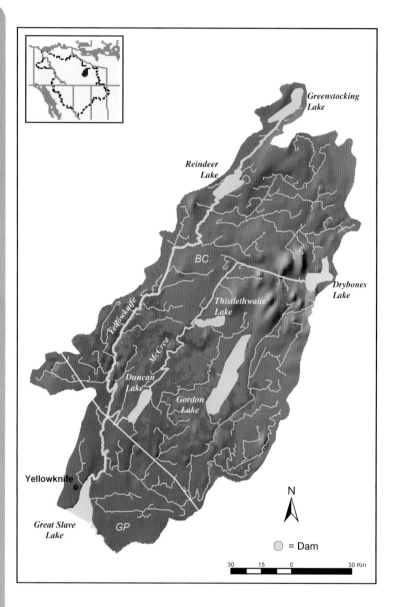

Relief: 706 m
Basin area: 11,300 km²
Mean discharge: 39 m³/s
Mean annual precipitation: 27 cm

Mean air temperature: −5.1°C
Mean water temperature: NA
No. of fish species: NA
No. of endangered species: NA

Physiographic provinces: Great Plains (GP), Bear–Slave–Churchill Uplands (BC)

Major fishes: northern pike, Arctic grayling, lake trout

Major other aquatic vertebrates: moose, beaver, muskrat, river otter, bald eagle, red-breasted merganser, common loon, northern pintail

Major benthic insects: NA

Nonnative species: none

Major riparian plants: black spruce, birch, poplar, willow, alder, Labrador tea, cranberry, bunchberry, bearberry

Special features: wilderness river draining northern region of Canadian Shield; small rivers connected by many lakes; wintering grounds for Bathurst caribou herd (north section of basin); gold mining

Fragmentation: one dam near mouth

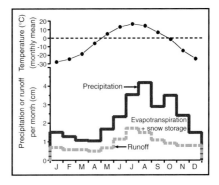

Yellowknife River near outlet of Lower Carp Lake, Northwest Territories (photo by C. Spence).

Chapter 19

Nelson and Churchill River Basins

David M. Rosenberg, Patricia A. Chambers, Joseph M. Culp,
William G. Franzin, Patrick A. Nelson, Alex G. Salki, Michael P. Stainton,
R. A. Bodaly, and Robert W. Newbury

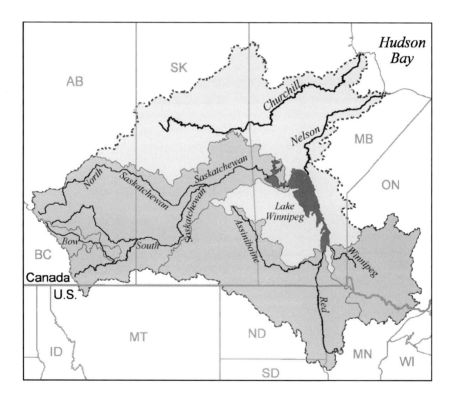

Two great Canadian rivers, the Nelson and the Churchill, drain waters mainly from the interior of Canada, cut through the Canadian Shield of northern Manitoba, and empty into Hudson Bay (see map). The two main-stem rivers come close together in northern Manitoba and were connected by way of Southern Indian Lake (Churchill) and the Rat-Burntwood River (Nelson) in the mid 1970s diverting 75% of the Churchill flow into the lower Nelson to augment flow for hydropower. Waters of the Nelson system begin their journey from the west on the eastern slopes of the Rocky Mountains, as two branches of the Saskatchewan River cross three Canadian provinces, and finally empty into Lake Winnipeg. Waters from the east originate in northwestern Ontario and flow via the Winnipeg River into the southern part of Lake Winnipeg. Waters from the south drain parts of Minnesota and North Dakota via the Red River of the North and also empty into the southern part of Lake Winnipeg. The Assiniboine River flows from the Canadian west into the Red River. The Nelson River proper originates at the outflow of Lake Winnipeg and carries its continental collection of water to Hudson Bay. In total, the Nelson system covers >1,000,000 km^2, and discharges 2480 m^3/s.

The Nelson River is of considerable historical importance in the development and settlement of the interior of Canada. Although European exploration began in the early 17th century, aboriginal peoples had already occupied the area for thousands of years. At least 39 different aboriginal groups have occupied the area of the lower Nelson and lower Churchill rivers since the glaciers retreated. Thomas Button, an Englishman, landed near the mouth of the Nelson River in 1612, two years after Europeans had begun exploring Hudson Bay. European explorers searching for the Northwest Passage to India had mapped the estuaries of the Seal, Churchill, Nelson, and Hayes rivers on Hudson Bay by the early 1600s. Button named the Nelson River after his sailing master, Frances Nelson of the HMS Resolution. Exploration of the region was begun in earnest in the late 17th century by the Hudson Bay Company, which established a post on the Nelson River in 1670, and on the neighboring Churchill River in 1688. The Hudson Bay Company (still today a major Canadian retailer) was formed in 1670 and established trading depots at the mouths of all the major rivers.

In this chapter, we describe the main-stem Nelson, as well as its 3 major tributaries (all >200 m^3/s), the Saskatchewan, Winnipeg, and Red River of the North. Also described are the Assiniboine and Bow rivers.

Nelson River

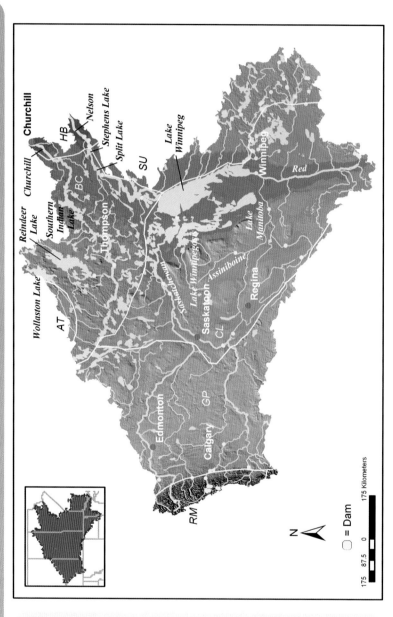

Relief: 3370 m (entire basin)
Basin area: 1,093,442 km²
(entire basin)
Mean discharge: 2480 m³/s
Mean annual precipitation: 52 cm

Mean air temperature: −3.4°C
Mean water temperature: NA
No. of fish species: 46 (main
stem); >94 (entire basin)
No. of endangered species: 0

Physiographic provinces: Bear–Slave–Churchill Uplands (BC), Hudson Bay Lowland (HB) (main stem only); five other provinces for entire basin

Major fishes: lake sturgeon, northern pike, brook trout, lake cisco, lake whitefish, longnose sucker, white sucker, burbot, walleye

Major other aquatic vertebrates: beaver, muskrat, mink, beluga whale, bearded seal

Major benthic insects: true flies (chironomid midges)

Nonnative species: NA (for main stem)

Major riparian plants: black spruce, tamarack, willow, alder, swamp birch, paper birch, trembling aspen, white spruce

Special features: on main stem, ice-caused "top hat" appearance of islands and "trimmed" shorelines in some reaches; pictographic sites throughout the region

Fragmentation: 5 hydropower dams on main stem; sites identified for future dams

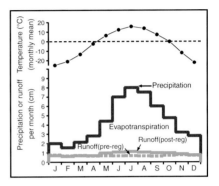

Whitemud Falls on Upper Nelson River (photo by R. Newbury).

Saskatchewan River

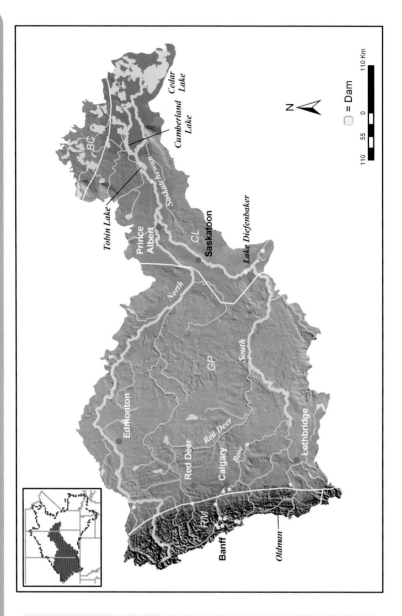

Relief: 3307 m
Basin area: 335,900 km²
Mean discharge: 567 m³/s
(postregulation)
Mean annual precipitation: 45 cm

Mean air temperature: −0.3°C
Mean water temperature: 9.7°C
No. of fish species: ≥48
No. of endangered species: 0

Physiographic provinces: Rocky
Mountains in Canada (RM), Great
Plains (GP), Central Lowland (CL),
Bear–Slave–Churchill Uplands (BC)
Major fishes: cutthroat trout, bull trout,
mountain whitefish, longnose sucker,
longnose dace, northern pike, walleye,
goldeye, yellow perch, quillback,
shorthead redhorse, lake sturgeon
Major other aquatic vertebrates:
beaver, mink, white pelican, river otter,
muskrat, tundra swan, ring-necked
duck
Major benthic insects: mayflies
(*Baetisca, Baetis, Ephemera,
Ephemerella, Ephoron, Heptagenia,
Tricorythodes*), stoneflies
(*Isoperla, Choroterpes*), caddisflies
(*Brachycercus, Cheumatopsyche,
Helicopsyche, Symphitopsyche,
Traverella*)
Nonnative species: brown trout,
rainbow trout, brook trout, purple
loosestrife, curly pondweed

Major riparian plants: red-osier
dogwood, sandbar willows, poplar,
water birch
Special features: originates in glaciers
and snowfields of Rocky Mountains,
a World Heritage Site; headwaters of
North Saskatchewan River in Banff
National Park designated Canadian
Heritage Rivers
Fragmentation: dams throughout for
hydropower and irrigation

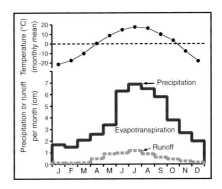

Saskatchewan River near Prince Albert, Saskatchewan (photo by L. Tebay).

Red River of the North

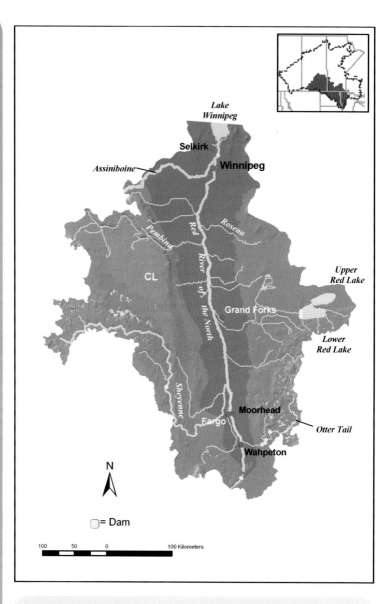

Lake Winnipeg

Selkirk

Winnipeg

Assiniboine

Pembina

Red

Roseau

River of the North

CL

Upper Red Lake

Grand Forks

Lower Red Lake

Sheyenne

Moorhead

Fargo

Otter Tail

Wahpeton

N

◯ = Dam

100   50   0   100 Kilometers

Relief: 350 m (includes
Assiniboine)
Basin area: 287,500 km²
(includes Assiniboine)
Mean discharge: 236 m³/s
(includes Assiniboine)
Mean annu. precipitation: 49 cm

Mean air temp.: 2.4°C
Mean water temp.: NA
No. of fish species: ~94
No. of endangered species: 14

Physiographic province: Central Lowland (CL)

Major fishes: channel catfish, black bullhead, walleye, sauger, freshwater drum, common carp, white sucker, shorthead redhorse, goldeye, mooneye, silver chub, emerald shiner, black crappie

Major other aquatic vertebrates: beaver, muskrat, western painted turtle, common snapping turtle, wood frog, chorus frog, spring peeper, northern leopard frog, gray treefrog, American toad, tiger salamander, blue spotted salamander, mud puppy

Major benthic insects: mayflies (*Baetis, Heptagenia, Ephoron, Hexagenia, Pentagenia, Isonychia, Tricorythodes*), caddisflies (*Ceratopsyche, Hydropsyche, Brachycentrus*), stoneflies (*Acroneuria, Pteronarcys*), true flies (*Tipula, Bezzia, Axarus*)

Nonnative species: common carp, white bass, largemouth bass, smallmouth bass

Major riparian plants: cottonwood, green ash, peach-leaved willow, burr oak, basswood, elm

Special features: very low-gradient large river

Fragmentation: main stem largely continuous; all major tributaries dammed

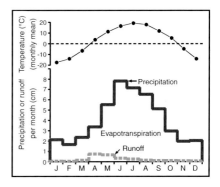

Red River of the North during the 1997 "Flood of the Century" at Emerson, Manitoba (photo by B. Oswald, Manitoba Department of Water Stewardship, Winnipeg).

Assiniboine River

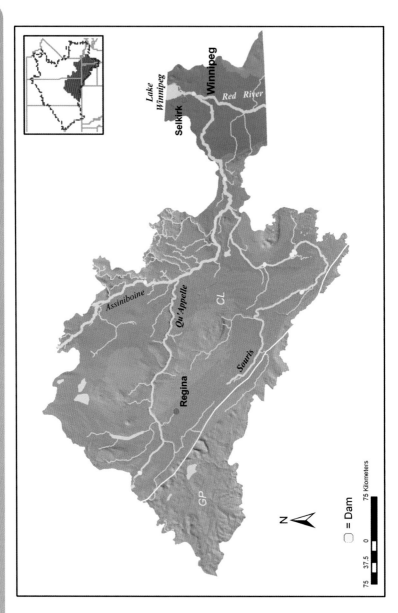

Relief: 350 m
Basin area: 162,000 km²
Mean discharge: 47.4 m³/s
Mean annual precipitation: 45 cm

Mean air temperature: 2.4°C
Mean water temperature: 9.0°C
No. of fish species: 55
No. of endangered species: 0

Physiographic province: Central Lowland (CL), Great Plains (GP)

Major fishes: walleye, sauger, channel catfish, goldeye, mooneye, common carp, white sucker, silver redhorse, golden redhorse, shorthead redhorse, quillback

Major other aquatic vertebrates: beaver, muskrat, wood frog, western chorus frog, spring peeper, northern leopard frog, gray treefrog, American toad, snapping turtle

Major benthic insects: mayflies (*Baetis, Heptagenia, Ephoron, Hexagenia, Pentagenia, Isonychia, Tricorythodes*), caddisflies (*Ceratopsyche, Hydropsyche, Brachycentrus*), stoneflies (*Acroneuria, Pteronarcys*), true flies (*Tipula, Bezzia*)

Nonnative species: common carp, white bass

Major riparian plants: several grasses, red-osier dogwood, peach-leaved willow, American elm, cottonwood, green ash, basswood, cattail

Special features: very long, relatively intact prairie river

Fragmentation: several dams on tributaries; 2 dams on main stem

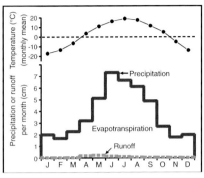

Assiniboine River near Winnipeg, Manitoba (photo by D. Moug).

Winnipeg River

○ = Dam

70   35   0   70 Km

N

Relief: 195 m   Mean air temperature: 2.4°C
Basin area: 135,800 km²   Mean water temperature: NA
Mean discharge: 850 m³/s   No. of fish species: 69
Mean annual precipitation: 62 cm   No. of endangered species: 4

Physiographic provinces: Central Lowland (CL), Superior Upland (SU)

Major fishes: northern pike, walleye, sauger, yellow perch, lake whitefish, lake trout, muskellunge, brook trout, smallmouth bass, white sucker, rainbow smelt

Major other aquatic vertebrates: beaver, river otter, muskrat, mink, loon, Canada goose, mallard, wood duck, great blue heron, herring gull, painted turtle

Major benthic insects: mayflies (*Caenis, Hexagenia*), caddisflies (*Hydropsyche, Cheumatopsyche*), true flies (*Chaoborus, Bezzia, Procladius, Cricotopus, Psectrocladius*)

Nonnative species: rainbow smelt, *Eubosmina coregoni, Bythotrephes cederstroemi*

Major riparian plants: bunchberry, Canada mayflower, chokecherry, dogwood, fragrant bedstraw, green

alder, highbush cranberry, lady fern, lowbush cranberry, mountain maple, northern bluebell, oak fern, pin cherry, pussy willow, rattlesnake plantain, slender wood grass, rose twisted stalk, sensitive fern, skunk currant, snowberry, starflower, white cedar

Special features: pristine wilderness except for dams; white-water rivers

Fragmentation: 33 major dams and control structures, many for hydropower

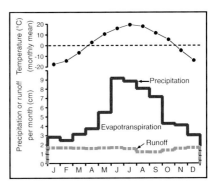

Winnipeg River below the White Dog Falls Station, Manitoba (photo by D. Watkinson).

Bow River

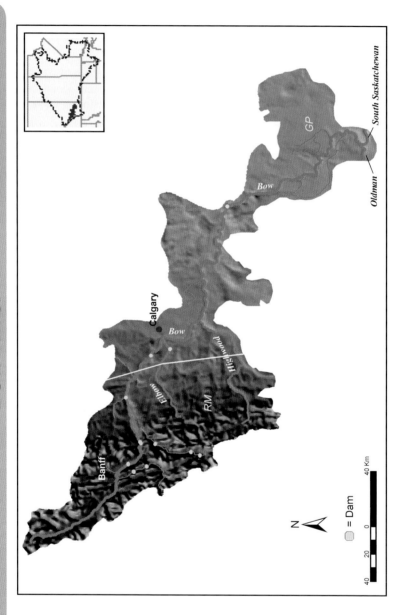

Relief: 2803 m
Basin area: 26,200 km²
Mean discharge: 91 m³/s
Mean annual precipitation: 40 cm

Mean air temperature: 3.9°C
Mean water temperature: 8.0°C
No. of fish species: 29
No. of endangered species: NA

Physiographic provinces: Rocky Mountains in Canada (RM), Great Plains (GP)

Major fishes: mountain whitefish, brook trout, brown trout, rainbow trout, bull trout, cutthroat trout, burbot, lake whitefish, northern pike, walleye, sauger, yellow perch, lake sturgeon, goldeye, mooneye, lake chub, emerald shiner, river shiner, spottail shiner, pearl dace, northern redbelly dace, quillback, longnose sucker, common white sucker, shorthead redhorse

Major other aquatic vertebrates: muskrat, beaver, mink, white pelican, Canada goose, bufflehead, common goldeneye, western grebe, gadwall, green-winged teal, horned grebe, lesser scaup, mallard, northern shoveler, pie-billed grebe, redhead

Major benthic insects: mayflies (*Baetis*, *Cinygmula*, *Ephemerella*, *Epeorus*, *Heptagenia*, *Isonychia*, *Rhithrogena*), caddisflies (*Cheumatopsyche*, *Glossosoma*, *Rhyacophila*, *Symphitopsyche*), stoneflies (*Capnia*, *Isoperla*, *Taenionema*, *Zapada*), true flies (Chironominae, Orthocladiinae, *Pericoma*)

Nonnative species: brook trout, brown trout, rainbow trout

Major riparian plants: river alder, willow, poplar, water birch, red-osier dogwood, balsam poplar, sandbar willow

Special features: originates in glaciers and snowfields

of Rocky Mountains in Alberta; World Heritage Site

Fragmentation: dams in mountains (for hydropower) and grasslands (for irrigation)

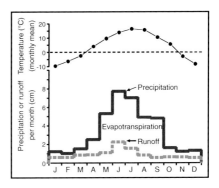

Bow River downstream of Lake Louise, Alberta (photo by P. Chambers).

Chapter 20

Rivers of Arctic North America

Alexander M. Milner, Mark W. Oswood, and Kelly R. Munkittrick

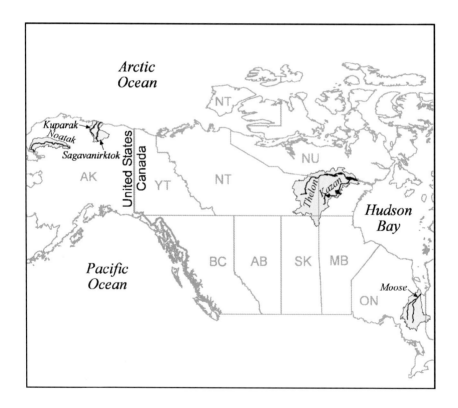

DOI: 10.1016/B978-0-12-375088-4.00020-8

This chapter covers rivers that lie within the Arctic region of North America, a vast area that encompasses northern regions of Alaska and Canada (see map). The largest Arctic rivers (Mackenzie, Nelson-Churchill and Yukon) are treated in chapters elsewhere. Although the Arctic is sometimes delineated by the Arctic Circle (66°N 32′W), this definition does not accurately reflect the characteristics of the region due to the overriding influence on climate of ocean currents and land mass topography. Consequently, the Arctic is frequently defined as regions north of the treeline, where mean July temperature does not exceed 10°C and at least one month is <0°C. There is a transitional zone (ecotone) between the treeless arctic tundra and the continuous closed canopy woodlands of the boreal forest, including the subarctic, where white and black spruce dominate, and mean monthly air temperature does not exceed 10°C for more than 4 months of the year and at least one month is <0°C. Thus, the coverage of the Arctic region in this chapter includes Hudson Bay in Canada, which extends as far south as 51° 20′, although the southerly part of the Bay is more accurately considered the subarctic. As Hudson Bay is frozen for the greater part of the year, the Arctic front of cold air masses extend further south in Canada than in Alaska due to the presence of this inland sea.

Most archeologists believe that the migration of humans into the New World occurred at least 15,000 years ago during the last Ice Age when the existence of the Bering Land Bridge created a connection between Alaska and Siberia termed Beringia.[1] However these Palaeo-Indians did not remain in Alaska but migrated southward to North, Central, and South America. When the Palaeo-Eskimos (Old Eskimos) arrived across the Bering Straits from the Chukchi Peninsula of Siberia they discovered the last major region on earth unoccupied by humans. The first discoveries of settlements date back 3000 to 4000 years, but they may have arrived earlier. The Northern Eskimos, the Inuit of Alaska and Canada, arrived more recently and descended from inhabitants originally in Alaska.

In this chapter, we describe 4 Arctic river systems as well as the Moose River, which is more correctly considered a Boreal Forest basin and flows into the south end of Hudson Bay. The Noatak, Kuparuk, and Thelon/Kazan are all pristine unimpaired waters, the largest being the Thelon/Kazan with a mean discharge of 1380 m^3/s. The Sagavanirktok River is potentially influenced by oil development, and the Moose (1370 m^3/s) is strongly influenced by hydro-electric dams, mining and forestry.

---

1 Langdon, S. 1989. "The native people of Alaska." 2nd ed. Greatland Graphics, Anchorage, Alaska.

Noatak River

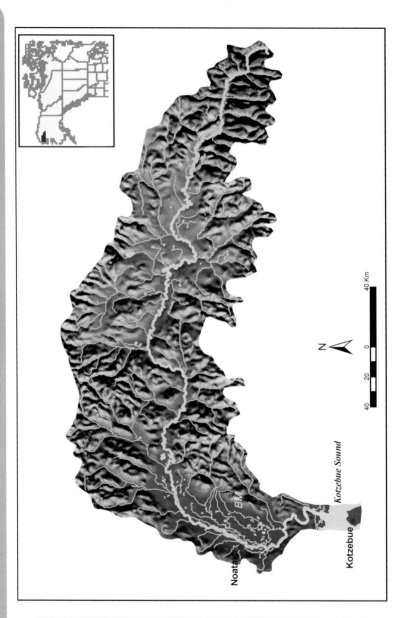

Relief: 2612 m
Basin area: 32,626 km^2
Mean discharge: 469 m^3/s
Mean annual precipitation: 33 cm

Mean air temperature: −5.8°C
Mean water temperature: NA
No. of fish species: 15 to 18
No. of endangered species: 0

Physiographic province: Brooks Range (BM)

Major fishes: chum salmon, Arctic char, Arctic grayling, humpback whitefish, round whitefish, least cisco, ninespine stickleback, slimy sculpin

Major other aquatic vertebrates: river otter, grizzly bear, loons, ducks, belted kingfisher

Major benthic insects: mayflies (Baetidae, Heptageniidae), true flies (chironomid midges), stoneflies (Chloroperlidae)

Nonnative species: none

Major riparian plants: willows, cottongrass, sedges

Special features: pristine Arctic tundra river within Gates of the Arctic National Park and Preserve and Noatak National Preserve; longest river segment in U.S. Wild and Scenic River system; little or no winter flow under deep ice cover; subsistence use of riverine resources by Inupiat Eskimos

Fragmentation: none

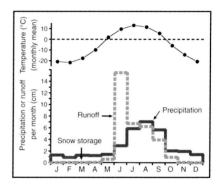

Noatak River, Alaska (photo by A. Milner).

Kuparuk River

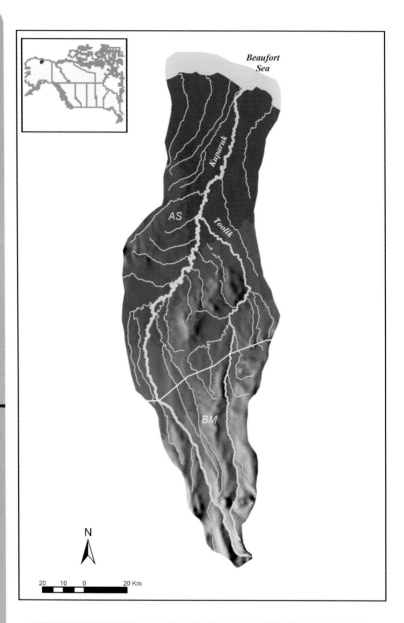

Relief: 1464 m
Basin area: 8107 km^2
Mean discharge: 40 m^3/s
Mean annual precipitation: 22 cm

Mean air temperature: −9.8°C
Mean water temperature: 2.6°C
No. of fish species: ≥6
No. of endangered species: 0

Physiographic provinces: Arctic Slope (AS), Brooks Range (BM)

Major fishes: Arctic grayling, broad whitefish, ninespine stickleback, slimy sculpin, burbot

Major other aquatic vertebrates: loons, swans, many duck species, including spectacled eider and king eider

Major benthic insects: true flies (*Orthocladius*, *Prosimulium*), mayflies (*Baetis*), caddisflies (*Brachycentrus*)

Nonnative species: none

Major riparian plants: dwarf willow, birch

Special features: pristine tundra river originating in the foothills of Brooks Range and traversing Arctic Coastal Plain to Beaufort Sea

Fragmentation: none

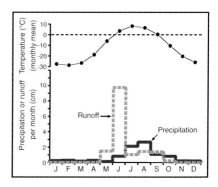

Kuparuk River, Alaska (photo by J. Benstead).

Sagavanirktok River

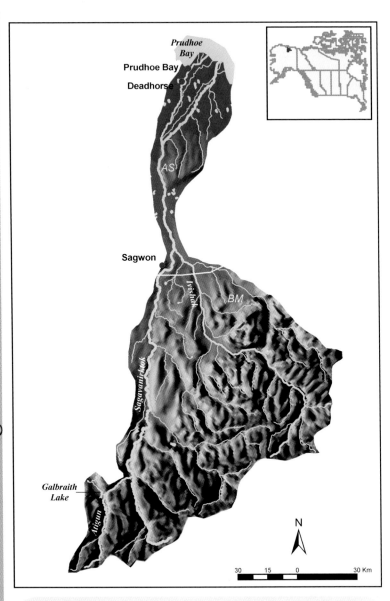

Relief: 2400 m
Basin area: 14,890 km^2
Mean discharge: 132 m^3/s
Mean annual precipitation: 11 cm
(underestimate)

Mean air temperature: −7.7°C
Mean water temperature: NA
No. of fish species: 10
No. of endangered species: 0

Physiographic provinces: Brooks
Range (BM), Arctic Slope (AS)

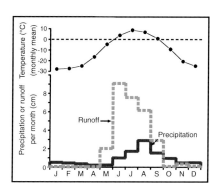

Major fishes: Dolly Varden, Arctic
grayling, Arctic cisco, broad whitefish,
round whitefish, lake trout

Major other aquatic vertebrates:
grizzly bear, spectacled eider, king
eider, yellow-billed loon

Major benthic insects: stoneflies
(*Capnia, Nemoura, Isogenus*),
mayflies (*Baetis, Cinygmula,
Ameletus*), true flies
(Orthocladiinae,
black flies)

Nonnative species:
none

Major riparian
plants: dwarf
willow, birch

Special features:
one of major
rivers on Alaskan
Arctic Slope
supporting large
fish populations;
unregulated;
delta provides
important bird
and fish habitat;
oil development,
including Trans-
Alaskan Pipeline,
along most of
river's length

Fragmentation:
none

Sagavanirktok River, Alaska, as it flows into Arctic Ocean
(photo by C. White).

Moose River

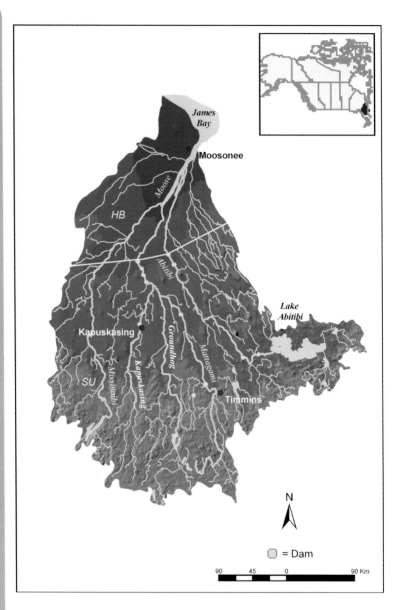

Relief: 325 m
Basin area: 109,000 km^2
Mean discharge: 1370 m^3/s
Mean annual precipitation: 80 cm

Mean air temperature: 0.1°C
Mean water temperature: 7.0°C
No. of fish species: 40
No. of endangered species: 0

Physiographic provinces: Superior Upland (SU), Hudson Bay Lowland (HB)

Major fishes: lake sturgeon, northern pike, walleye, white sucker, troutperch, pearl dace, longnose dace, spotfin shiner, emerald shiner

Major other aquatic vertebrates: muskrat, beaver, fisher, marten, mink, river otter, moose, lesser snow goose

Major benthic insects: mayflies (*Hexagenia, Baetis, Caenis, Ephemerella, Heptagenia, Stenonema*), true flies (*Chironomus, Paratanytarsus, Cladopelma*), stoneflies (Perlodidae, Pteronarcyidae)

Nonnative species: none

Major riparian plants: white birch, trembling aspen, white spruce, black

spruce, jack pine, dogwood, alder, willow, grasses, sedges, rushes, horsetails, arrowheads, pondweed, coontail

Special features: >200 waterfalls and the Missinaibi pristine Heritage River

Fragmentation: more than 40 dams and water-control structures throughout basin

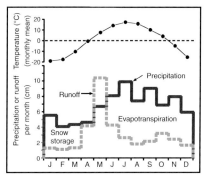

The confluence of two tributaries of the Moose River, Ontario (photo by W. Gibbons).

Thelon/Kazan River

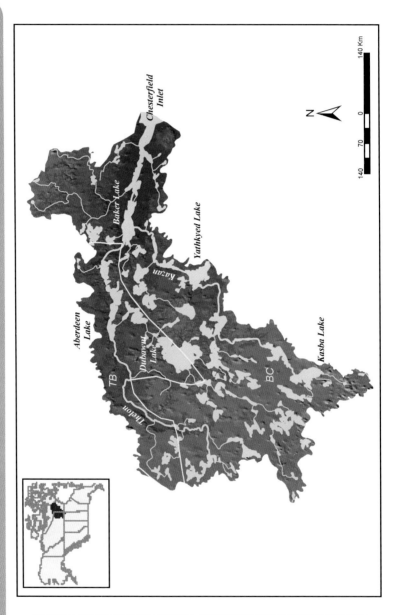

Relief: 300 to 500 m
Basin area: 239,332 km²
Mean discharge: 1380 m³/s
Mean annual precipitation: 17 cm

Mean air temperature: −8.7°C
Mean water temperature: NA
No. of fish species: 13
No. of endangered species: 0

Physiographic provinces: Thelon Plains and Bear River Lowland (TB), Bear–Slave–Churchill Uplands (BC)

Major fishes: lake trout, Arctic grayling, northern pike, Arctic char, humpback whitefish, round whitefish, cisco, slimy sculpin, spoonhead culpin, lake chub

Major other aquatic vertebrates: grizzly bear, bald eagle, tundra swan, Canada goose

Benthic insects: NA

Nonnative species: none

Major riparian plants: willow shrub, bog birch, birch, spruce

Special features: Canadian Heritage River (both Thelon and Kazan); large remote wilderness river; Thelon Game Sanctuary; Kazan Falls

Fragmentation: NA

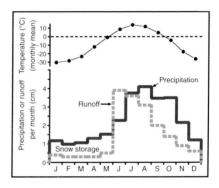

Thelon River, "oasis" section below Warden's Grove Northwest Territories (photo by C. Hayne).

Chapter 21

Atlantic Coast Rivers of Canada

Richard A. Cunjak and Robert W. Newbury

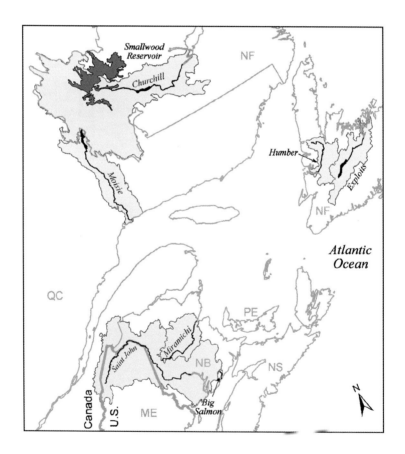

The rivers of Eastern Canada flow into the Atlantic Ocean from Newfoundland, Labrador, Nova Scotia, Prince Edward Island, New Brunswick, and parts of eastern Quebec. Geographically, the region is bounded by the Atlantic Ocean on the south and east, 70°W longitude on the west, and 55°N latitude in the north. The northern boundary generally follows a height of land between rivers flowing north to Ungava Bay or south into the Gulf of St. Lawrence (see map). To the north, the rivers run through near wilderness regions of the rocky Canadian Shield. To the south, many of the rivers flow through the Atlantic lowlands in wide valleys formed by large glacial meltwater rivers. Towns, industries, and farmlands now occupy the valley flats and shallow plains beside the smaller present-day rivers. To the west, the interior drainage of Eastern Canada (the Great Lakes and various downstream tributaries) flows into the Atlantic through the St. Lawrence River, the second largest river (by discharge) in North America discussed in Chapter 22. The Atlantic salmon, a symbol of economic value and environmental quality, is native to most rivers in the region. The state of these rivers today can be assessed by understanding the state of their salmon.

The rivers have been used by aboriginal peoples for over 9000 years—the Inuit along the coast of Labrador, the Innu in central and southern Labrador, the Montagnais and Mi'kmaqs of the Maritimes, and the now extinct Beothuks of insular Newfoundland. These rivers have long served these peoples for movement between seasonal settlements, and for accessing hunting and fishing areas. Travel to eastern Canada by Europeans includes some of the earliest explorers. Basques traveled regularly to southern Labrador between 1530 and 1600, primarily in search of whales and cod; Norwegian sailors are known to have reached northwestern Newfoundland before the 10th Century. Colonization of eastern Canada, primarily by French and English explorers such as Jacques Cartier and John Cabot, began in the early 16th Century, mainly along the coast of the Bay of Fundy and along the St. Lawrence River, the route to the interior of North America.

In this chapter, we describe the Moisie and Churchill rivers north of the Gulf of St. Lawrence, the Humber and Exploits rivers on insular Newfoundland, and the Miramichi, Big Salmon, and St. John in the Maritime Provinces. The St. John (draining parts of Maine, Quebec, and New Brunswick) and the Churchill (Labrador) are easily the largest of these rivers with discharges $>1000 \, m^3/s$. The Exploits, Miramichi, Moisie, and Humber rivers are relatively large ($>200 \, m^3/s$), and the Big Salmon is much smaller.

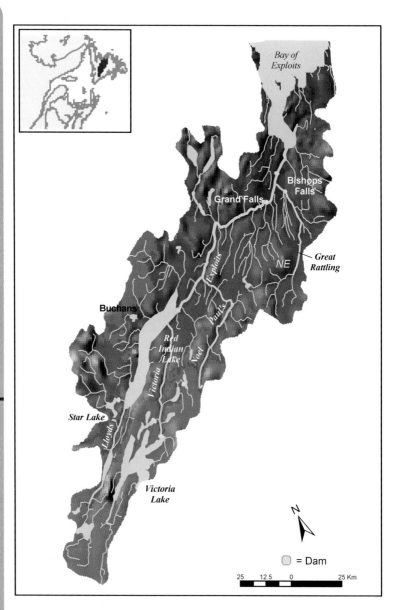

Exploits River

Relief: 490 m
Basin area: 11,272 km²
Mean discharge: 270 m³/s
Mean annual precipitation: 99 cm

Mean air temperature: 4.4°C
Mean water temperature: NA
No. of fish species: 6
No. of endangered species: 0

Physiographic province: New England/Maritime (NE)

Major fishes: Atlantic salmon, brook trout, threespine stickleback, rainbow smelt, American eel, Arctic char

Major other aquatic vertebrates: beaver, river otter, mink

Major benthic insects: true flies (chironomid midges)

Nonnative species: NA

Major riparian plants: black spruce, balsam fir, speckled alder, bog laurel, Labrador tea

Special features: large island river with second largest lake in Newfoundland (Red Indian Lake); historic meeting place of Europeans and last of Beothuk peoples at Red Indian Lake; major

midwinter ice jams below storage dams caused instant and catastrophic flooding in riverside towns.

Fragmentation: nine hydroelectric and water-storage dams in basin, some with no fish-passage facilities (e.g., Star Lake); water diversion from Victoria River to adjacent river basin

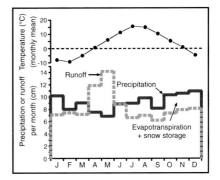

Exploits River and hydroelectric dam at Bishop's Falls, Newfoundland (photo by C. Bourgeois).

Miramichi River

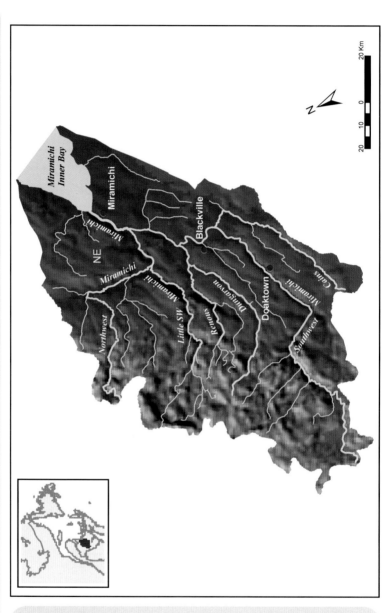

Relief: 764 m
Basin area: 14,000 km^2
Mean discharge: 322 m^3/s
Mean annual precipitation:
112 cm

Mean air temperature: 4.7°C
Mean water temperature: NA
No. of fish species: 21
freshwater, 8 diadromous
No. of endangered species: 0

Physiographic province: New England/Maritime (NE)

Major fishes: Atlantic salmon, brook trout, sea lamprey, American eel, alewife, American shad, rainbow smelt, striped bass, Atlantic sturgeon, slimy sculpin, blacknose dace, common shiner, lake chub, white sucker, sea lamprey

Major other aquatic vertebrates: beaver, river otter, mink, northern two-lined salamander, double-crested cormorant

Major benthic insects: mayflies (*Baetis, Ephemerella, Stenonema*), stoneflies (*Leuctra, Alloperla, Pteronarcys*), caddisflies (*Hydropsyche, Dolophilodes, Glossosoma, Pycnopsyche, Brachycentridae*), true flies (*Prosimulium, Tipula*, chironomid midges)

Nonnative species: white perch, brown trout, chain pickerel

Major riparian plants: white spruce, white pine, black spruce, yellow birch, northern red oak, sweet gale, speckled alder

Special features: several principal branches and tributaries set in broad

glacial river valley; produces more Atlantic salmon than any other river in North America; very popular for anglers and canoeists

Fragmentation: none

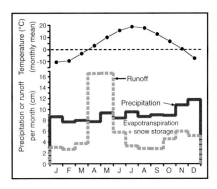

Southwest Miramichi River near Doaktown, New Brunswick (photo by R. Newbury).

St. John River

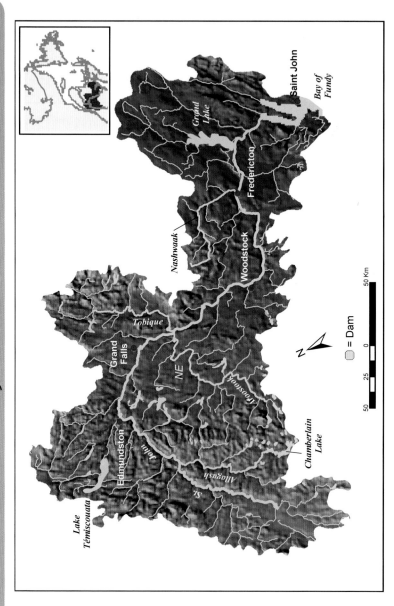

Relief: 820 m
Basin area: 55,110 km^2
Mean discharge: 1110 m^3/s
Mean annual precipitation: 114 cm

Mean air temperature: 5.3°C
Mean water temperature: NA
No. of fish species: 36
No. of endangered species: 3

Physiographic province: New England/Maritime (NE)

Major fishes: common shiner, blacknose dace, brown bullhead, slimy sculpin, white sucker, Atlantic salmon, brook trout, striped bass, American eel, alewife, Atlantic sturgeon, shortnose sturgeon, sea lamprey, yellow perch

Major other aquatic vertebrates: muskrat, beaver, common merganser, double-crested cormorant

Major benthic insects: true flies (chironomid midges, black flies), mayflies (Baetidae, Heptageniidae, Ephemerellidae), stoneflies (Chloroperlidae, Perlidae), caddisflies (Philopotamidae, Hydropsychidae)

Nonnative species: muskellunge, chain pickerel, brown trout, rainbow trout, smallmouth bass

Major riparian plants: white spruce, white pine, balsam poplar, willows, white birch, northern red oak, speckled alder

Special features: longest river in eastern Canada; international border; "Reversing Falls" near mouth due to large tidal amplitude from Bay of Fundy; extensive marshes in lower river; >100 km under tidal influence

Fragmentation: 11 hydroelectric dams from headwaters to mouth

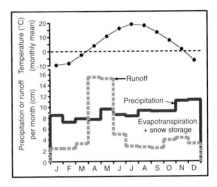

St. John River downstream from Moody Bridge in northern Maine (photo by M. Gautreau).

Moisie River

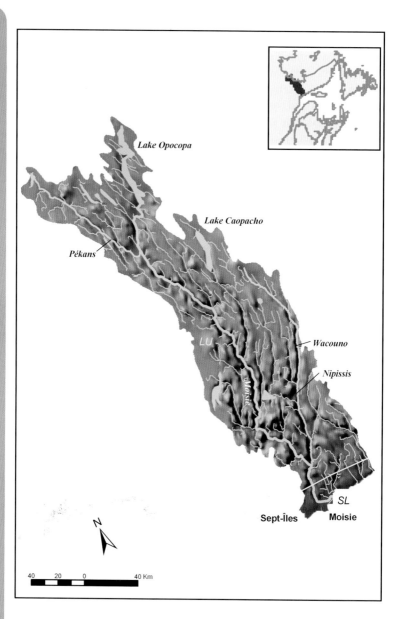

Relief: 1011 m
Basin area: 19,871 km^2
Mean discharge: 426 m^3/s
Mean annual precipitation: 116 cm

Mean air temperature: 1.0°C
Mean water temperature: 6.0°C
No. of fish species: ≥13
No. of endangered species: 0

Physiographic province: Laurentian Highlands (LU)

Major fishes: Atlantic salmon, brook trout, northern pike, white sucker, American eel, lake chub, lake whitefish, threespine stickleback, longnose sucker, burbot, sea lamprey

Major other aquatic vertebrates: beaver, mink, muskrat

Major benthic insects: stoneflies (*Paracapnia, Isoperla*, Leuctridae, Perlidae), mayflies (*Ephemerella, Heptagenia, Ameletus*), caddisflies (*Lepidostoma, Chimarra, Hydropsyche*), true flies (chironomid midges)

Nonnative species: none

Major riparian plants: black spruce, white spruce, balsam fir, white birch, speckled alder, trembling aspen, Labrador tea

Special features: exceptional scenery and wilderness setting; stock of large Atlantic salmon; traditional route to Quebec interior for trapping and hunting for Montagnais Indians; recent designation as aquatic reserve to ensure protection from development

Fragmentation: no dams, although past interest in development of hydroelectric dams and diversions

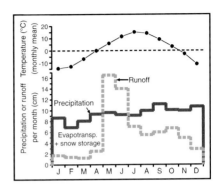

Katchapahum Falls on the Moisie River, Quebec (photo by K. Schiefer).

Big Salmon River

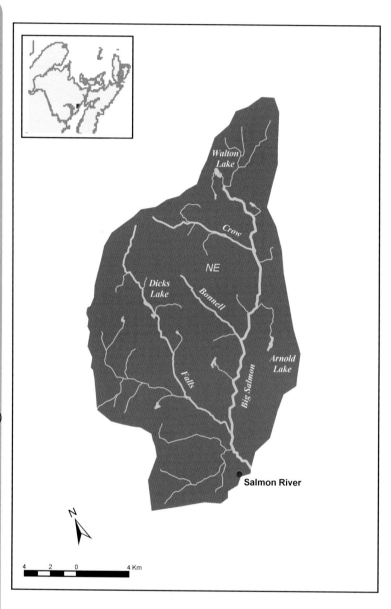

Relief: 410 m
Basin area: 332 km²
Mean discharge: 11.8 m³/s
Mean annual precipitation: 146 cm

Mean air temperature: 5.3°C
Mean water temperature: NA
No. of fish species: 7
No. of endangered species: 1

Physiographic province: New England/Maritime (NE)

Major fishes: Atlantic salmon, brook trout, American eel, blacknose dace

Major other aquatic vertebrates: mink, beaver

Major benthic insects: mayflies (Ephemerellidae, Baetidae), stoneflies (Leuctridae), true flies (chironomid midges, black flies)

Nonnative species: rainbow trout, Arctic char (Walton Lake)

Major riparian plants: red spruce, white pine, speckled alder, white birch

Special features: unique wilderness coastal basin of Bay of Fundy with pristine, old-growth forest in deeply incised river valley; historically one

of the most productive Inner Fundy salmon rivers, but today few salmon return

Fragmentation: dam near head of tide on main branch removed in 1963; currently small water-level control dams at outlets of several headwater lakes (e.g., Walton Lake)

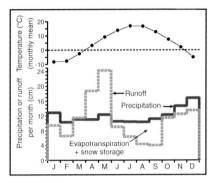

Lower reach of the Big Salmon River, New Brunswick (photo by R. Newbury).

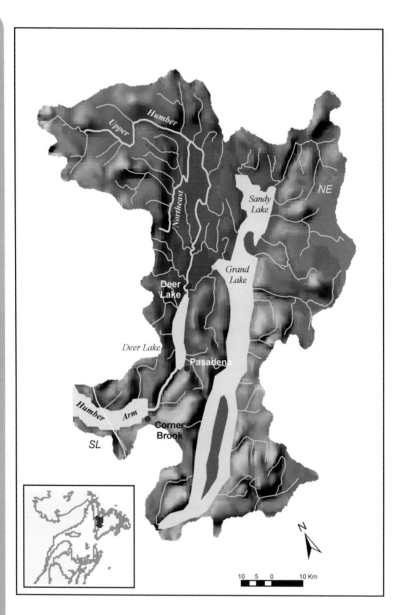

Relief: 700 m
Basin area: 7860 km^2
Mean discharge: 246 m^3/s
Mean annual precipitation: 119 cm

Mean air temperature: 5.2°C
Mean water temperature: NA
No. of fish species: 7
No. of endangered species: 0

Physiographic province: New England/Maritime (NE)

Major fishes: Atlantic salmon, brook trout, rainbow smelt, American eel, Arctic char, threespine stickleback

Major other aquatic vertebrates: mink, river otter, beaver

Major benthic insects: true flies (black flies)

Nonnative species: rainbow trout

Major riparian plants: black spruce, white birch, speckled alder, juniper, sweet gale, bog laurel

Special features: popular sport-fishing river for Atlantic salmon and brook trout; Grand Lake is largest lake in Newfoundland; landlocked populations of Atlantic salmon (ouananiche) and Arctic char

upstream of impassable falls; parts of headwaters in Gros Morne National Park and provincial park

Fragmentation: natural obstruction at Main Falls (6.4 m), 113 km from river mouth; North Brook has natural obstruction 14.8 km from mouth; Grand Lake used for hydro-power; several hydro storage dams in Grand Lake subbasin

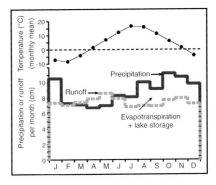

Lower Humber River, Shell Bird Island, Newfoundland (photo by D. Hall).

Churchill River

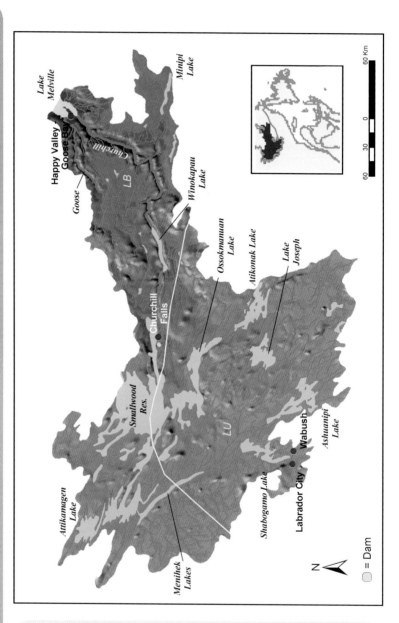

Relief: 549 m
Basin area: 93,415 km²
Mean discharge: 1861 m³/s
Mean annual precipitation: 95 cm

Mean air temperature: −3.5°C
Mean water temperature: NA
No. of fish species: 20
No. of endangered species: 0

Physiographic provinces: Labrador Highlands (LB), Laurentian Highlands (LU)

Major fishes: northern pike, lake whitefish, white sucker, brook trout, lake trout, Atlantic salmon (landlocked salmon only above Muskrat Falls), Arctic char, lake chub, threespine stickleback, mottled sculpin, slimy sculpin

Major other aquatic vertebrates: muskrat, beaver, mink

Major benthic insects: NA

Nonnative species: none

Major riparian plants: black spruce, white birch, larch, Labrador tea

Special features: second-largest river (by discharge) in Atlantic drainage; Muskrat Falls, an 8 m waterfall 40 km from mouth, is complete barrier to

upstream migration; original Innu name "Mishtashipu" means "big river"

Fragmentation: Churchill Falls Dam, huge hydroelectric project (5400 MW), a major energy provider for northeastern United States; many dams and dykes for hydroelectric project act as unnatural barriers to fish migration

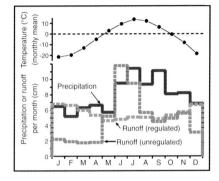

Churchill Falls on the Churchill River, Labrador, before diversion to power station in 1970.

Chapter 22

St. Lawrence River Basin

James H. Thorp, Gary A. Lamberti, and Andrew F. Casper

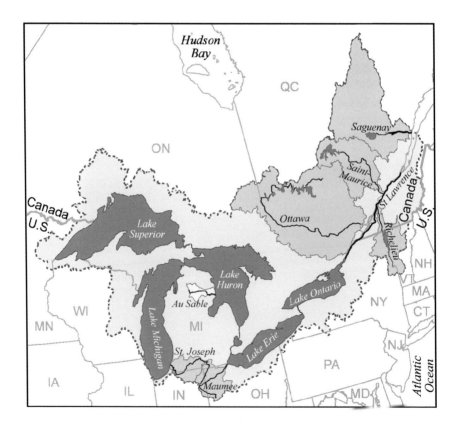

The second largest river network in North America in annual discharge is the international St. Lawrence River–Great Lakes System. The importance of this river-lake system in times of increasing global shortage of freshwater is hard to overestimate because its basin holds about 23,000 km³ of water (roughly 18% of the world's freshwaters).[1] The river's catchment stretches from ~40° to 50°N latitude and ~65° to 93°W longitude (see map). The St. Lawrence River–Great Lakes System forms part of the boundary between Canada and the United States, and in some places, it physically divides/links various Indian Nations, including the Mohawks in northern New York and southern Canada. With a watershed of ~1.6 million km², water in this system can travel at least 3260 km from western Lake Superior to the Cabot Strait in the estuarine Gulf of St. Lawrence. Along this lentic-lotic-estuarine pathway to the sea, the river system draws sustenance from eight states (Minnesota, Wisconsin, Illinois, Indiana, Michigan, Ohio, Pennsylvania, and New York) and at least two provinces (mostly Ontario and Québec).

For perhaps 8500 to 9000 years the St. Lawrence River–Great Lakes System has played an important role in the lives of many nations of Native Americans, with Abnake, Algonquin, Huron, Iroquois, Montagnais, Potawatomi, and other groups thriving in this area. One estimate of the sixteenth-century population of Native Americans around the Great Lakes alone was 60,000 to 117,000 people. The earliest surviving record of the basin's exploration by Europeans dates to 1535, during a period of exploration for the fabled Northwest Passage from the Atlantic to the Pacific. In that year, the French explorer Jacques Cartier happened upon this great river and named it in honor of the coincident feast day of Saint Lawrence. The first permanent European settlement in the St. Lawrence River valley was established by the French explorer Samuel de Champlain in the first decade of the seventeenth century near the present-day city of Québec. The river changed hands in 1763 following Britain's victory in the so-called French and Indian War.

Although a disproportionate amount of scientific knowledge exists about the lake portions of this enormous lake-river system (i.e., on the five major Great Lakes), this chapter describes characteristics of the main-stem St. Lawrence River, some of its major tributaries, and some tributaries that flow into the Great Lakes. We describe the Ottawa, Saint-Maurice, Saguenay, and Richelieu that flow into the St. Lawrence, the largest of which is the Ottawa River with a discharge of almost 2000 m³/s. The much smaller tributaries of the Great Lakes include the St. Joseph, Maumee, and AuSable.

---

1 Fuller, K., H. Shear, & J. Wittig (eds.). 1995. The Great Lakes: an environmental atlas and resource book, 3rd ed. U.S. Environmental Protection Agency, Great Lakes Environmental Program Office, Chicago; Government of Canada, Toronto, Ontario.

St. Lawrence River Main Stem

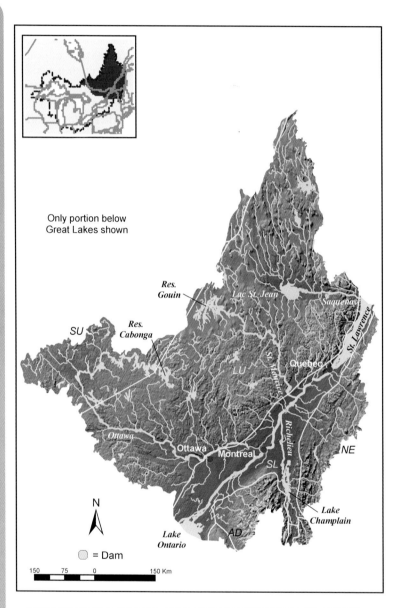

Only portion below
Great Lakes shown

Res.
Gouin

Lac St. Jean

Saguenay

SU

Res.
Cabonga

LU

St. Maurice

Quebec

St. Lawrence

Ottawa

Ottawa   Montreal

Richelieu

NE

SL

N

Lake
Champlain

Lake
Ontario

AD

⬤ = Dam

150   75   0   150 Km

Relief: 1945 m
Basin area: 574,000 km^2
(main stem)
Mean discharge: 12,600 m^3/s
(excluding Saguenay)
Mean annual precipitation: 94 cm

Mean air temperature: 6.7°C
Mean water temperature: 9.7°C
No. of fish species: 87 freshwater,
18 diadromous
No. of endangered species: ~20

Physiographic provinces: Laurentian Highlands (LU), Superior Uplands (SU), St. Lawrence Lowland (SL), Adirondack (AD), New England/Maritime (NE)

Major fishes: lamprey, lake sturgeon, Atlantic sturgeon, gar, bowfin, American eel, alewife, gizzard shad, creek chub, fallfish, yellow perch, walleye, white sucker, silver redhorse, channel catfish, tadpole madtom, muskellunge, central mudminnow, rainbow smelt, brown trout, troutperch, banded killifish, burbot, brook silverside, mottled sculpin

Major other aquatic vertebrates: muskrat, river otter, beaver, mink, beluga whale, long-finned pilot whale, blue-winged teal, gadwall, American wigeon, great blue heron, ring-billed gull, belted kingfisher, cormorant, tern, bald eagle, osprey, northern water snake, painted turtle

Major benthic insects: mayflies (*Stenonema, Hexagenia*), caddisflies (*Nectopsyche*), stoneflies (*Pteronarcys*)

Nonnative species: common carp, rainbow trout, white perch, *Bithynia*

tentaculata and *Viviparus georgianus* (snails), zebra mussel, quagga mussel, *Echinogammarus ischnus* (amphipod), purple loosestrife

Major riparian plants: silver maple, red maple, black ash, green ash, black willow, American basswood, cattails, sedges, rushes, bulrushes, reed canary grass

Special features: lower river continentally outstanding; RAMSAR World Heritage Site (Lac Saint-Pierre); least turbid of world's 15 largest rivers

Fragmentation: 4 hydroelectric dams and 7 navigation locks on main stem; many dams on tributaries

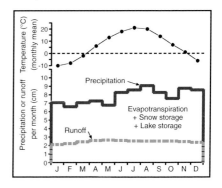

St. Lawrence River near Ivy Lea, Ontario (photo by J. M. Farrell, Thousands Islands Biological Station).

Ottawa River

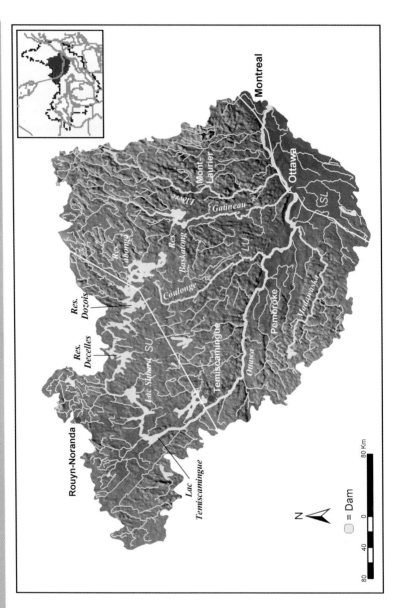

Relief: 911 m
Basin area: 146,334 km²
Mean discharge: 1948 m³/s
Mean annual precipitation: 100 cm

Mean air temperature: 6.0°C
Mean water temperature: 9.6°C
No. of fish species: 53
No. of endangered species: 12

Physiographic provinces: St. Lawrence Lowlands (SL), Laurentian Highlands (LU), Superior Upland (SU)

Major fishes: sturgeon, walleye, sauger, muskellunge, northern pike, yellow perch, crappie, lake whitefish, lake cisco, largemouth bass, smallmouth bass, channel catfish, brown bullhead, copper redhorse, suckers, cyprinids, mooneye

Major other aquatic vertebrates: muskrat, river otter, beaver, mink

Major benthic insects: true flies (*Polypedilum*), mayflies (*Hexagenia, Stenonema*), stoneflies (*Isoperla*), caddisflies (*Polycentropus, Brachycentrus*)

Nonnative species: brown trout, rainbow trout, zebra mussel, quagga mussel, purple loosestrife

Major riparian plants: wild rice, bur-reed, arrowhead, northern bugleweed, fox sedge, silverweed, marsh speedwell, calamus root, bulrushes, Small's spikerush, knotsheath sedge, red top, reed canary grass, silver maple, green ash, alder, willow

Special features: 8 provincial faunal reserves; island–wetland complex (Petri Islands Preserve) near confluence with St. Lawrence

Fragmentation: 7 dams on main stem; >300 dams on tributaries

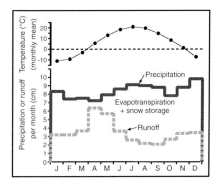

Ottawa River.

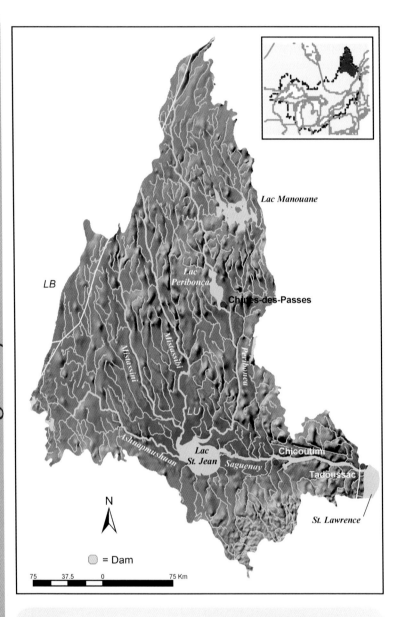

Saguenay River

Lac Manouane

LB

Lac
Peribonca

Chutes-des-Passes

Mistassini

Mistassibi

Peribonca

LU

Ashuapmushuan

Lac
St. Jean

Saguenay

Chicoutimi

Tadoussac

N

St. Lawrence

⬤ = Dam

75   37.5   0   75 Km

Relief: 1130 m
Basin area: 85,500 km²
Mean discharge: 1535 m³/s
Mean annual precipitation:
97 cm

Mean air temperature: 3.0°C
Mean water temperature: NA
No. of fish species: 76 in river
and fjord
No. of endangered species: 6

Physiographic provinces: Laurentian Highlands (LU), Labrador Highlands (LB)

Major fishes: Atlantic salmon, fallfish, brown bullhead, lake whitefish, brook trout, perch, burbot, rainbow smelt, lake (black) sturgeon, American eel, emerald shiner, sauger, white sucker, longnose sucker, northern pike

Major other aquatic vertebrates: beaver, river otter, muskrat, mink, long-finned pilot whale, beluga whale, great blue heron, black-crowned night heron; many bird species in fjord, including red knot, sanderling, purple sandpiper, peregrine falcon, osprey

Major benthic insects: NA

Nonnative species: rainbow trout, purple loosestrife and >20 other plants in riparian zone

Major riparian plants: sparse vegetation in upper third due to dams and log driving; middle third dominated by American bulrush but with ≥250 species; lower third with sparse patches of American bulrush and, where salinity intrudes, saltwater cord grass

Special features: large Lac Saint-Jean in headwaters; last 100 km flows through largest fjord in northwest Atlantic; continentally and globally distinct site (Batture aux Alouettes)

for migratory birds at confluence with St. Lawrence

Fragmentation: 3 hydroelectric/navigation dams on main stem and >300 small and medium-size dams in watershed

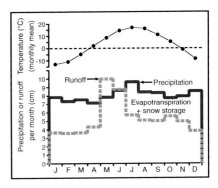

Saguenay River at Cap-Eternite, Quebec (photo by M. Rautio).

St. Joseph River

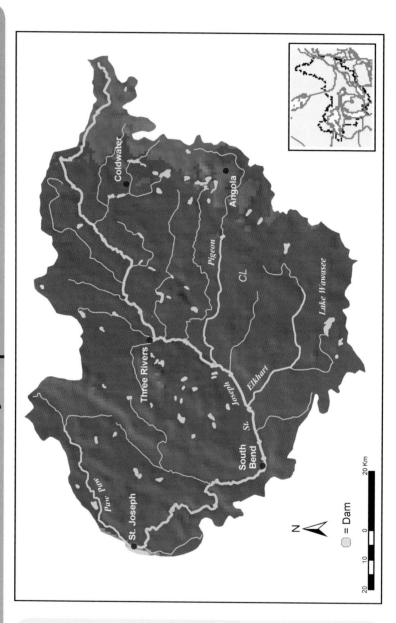

Relief: 200 m
Basin area: 12,150 km²
Mean discharge: 96.3 m³/s
Mean annual precipitation: 99 cm

Mean air temperature: 9.7°C
Mean water temperature: 12.0°C
No. of fish species: 114
No. of endangered species: 127

Physiographic province: Central Lowland (CL)

Major fishes: brook trout, smallmouth bass, bluegill, walleye, white sucker, hornyhead chub, creek chub, yellow perch, logperch, pirate perch, blacknose dace, blackside darter, rainbow darter, bluntnose minnow, common shiner, common stoneroller, central mudminnow, northern hogsucker, mottled sculpin, channel catfish, northern madtom

Major other aquatic vertebrates: beaver, river otter, muskrat, mink

Major benthic insects: mayflies (*Baetis*, *Hexagenia*), stoneflies (*Amphinemura*, *Paracapnia*, *Taeniopteryx*), caddisflies (*Cheumatopsyche*, *Glossosoma*)

Nonnative species: brown trout, rainbow trout, chinook salmon, coho salmon, sea lamprey, common carp, goldfish, zebra mussel, Asiatic clam, purple loosestrife, Eurasian watermilfoil

Major riparian plants: ash, elm, cottonwood, maple, oak, poplar, wild celery, pondweed

Fragmentation: 17 dams on main stem, 190 dams in basin

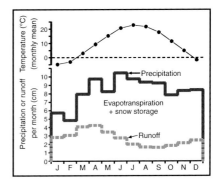

St. Joseph River near South Bend, Indiana (photo by G. Lamberti).

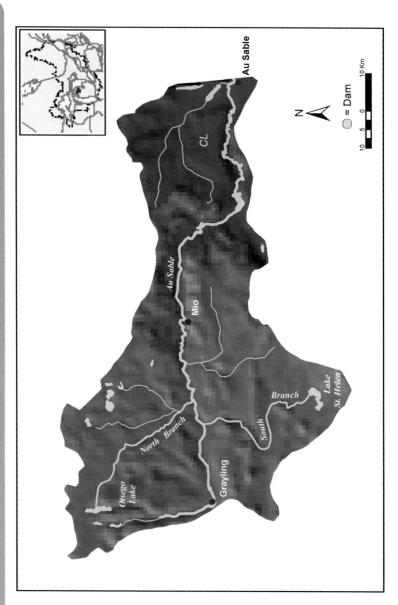

Ausable River

Relief: 250 m
Basin area: 5240 km²
Mean discharge: 42.1 m³/s
Mean annual precipitation: 83 cm

Mean air temperature: 5.5°C
Mean water temperature: 11.5°C
No. of fish species: 93
No. of endangered species: 12

Physiographic province: **Central Lowland (CL)**

Major fishes: brook trout, brown trout, rainbow trout, sculpins, shiners, white sucker, other suckers, several dace, Johnny darter

Major other aquatic vertebrates: muskrat, beaver, river otter, mink

Major benthic insects: true flies (*Antocha, Tipula*), mayflies (*Baetis, Ephemerella, Hexagenia, Rhithrogena, Stenonema*), stoneflies (*Paragnetina*), caddisflies (*Brachycentrus, Helicopsyche, Hydropsyche, Rhyacophila*)

Nonnative species: ≥7 fishes, including brown trout, rainbow trout, sea lamprey

Major riparian plants: alder, ash, cedar, fir, red maple, poplar, spruce, tamarack, willow

Special features: 37 km of main stem designated National Wild and Scenic River; 560 km (157 km of main stem) in State Natural Rivers Program; about one-third of watershed in Huron National Forest

Fragmentation: 67 dams along main stem and tributaries, ranging from <1 to >10 m in height

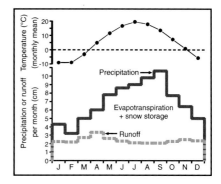

AuSable River near mouth, Michigan (photo by Tim Palmer).

Maumee River

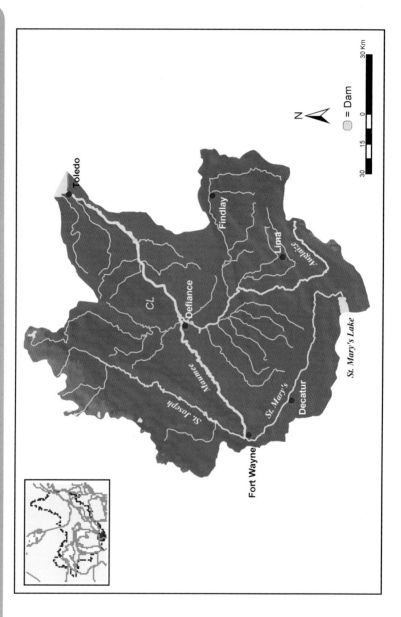

Relief: 73 m
Basin area: 16,458 km²
Mean discharge: 150 m³/s
Mean annual precipitation: 84 cm

Mean air temperature: 9.0°C
Mean water temperature: 12.7°C
No. of fish species: >60
No. of endangered species: 6

Physiographic province: Central Lowland (CL)

Major fishes: topminnow, chubs, spotfin shiner, bluntnose minnow, hognose sucker, redhorse sucker, blacknose dace, mudminnow, variegated darter, green sunfish, smallmouth bass, yellow bullhead, channel catfish, madtoms, walleye, crappie, gizzard shad, longnose gar, white perch, logperch, drum

Major other aquatic vertebrates: muskrat, beaver

Major benthic insects: true flies (*Rheotanytarsus*, *Simulium*), mayflies (*Baetis*), caddisflies (*Cheumatopsyche*)

Nonnative species: 11 fishes (goldfish, carp, sea lamprey, white perch, round goby), rusty crayfish, quagga mussel, zebra mussel, purple loosestrife

Major riparian plants: sycamores, black locust, beech, sugar maple

Special features: largest drainage area of any Great Lakes river; largest population of migrating walleye east of Mississippi River; 112 km as Ohio state scenic river; Lower Cedar Creek (Indiana) listed as outstanding segment

Fragmentation: 2 dams on main stem and 3 relatively large dams on tributaries

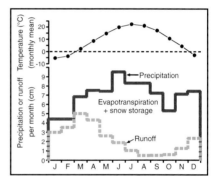

Maumee River in Grand Rapids, Ohio (photo by D. Ramsey).

Rivière Richelieu

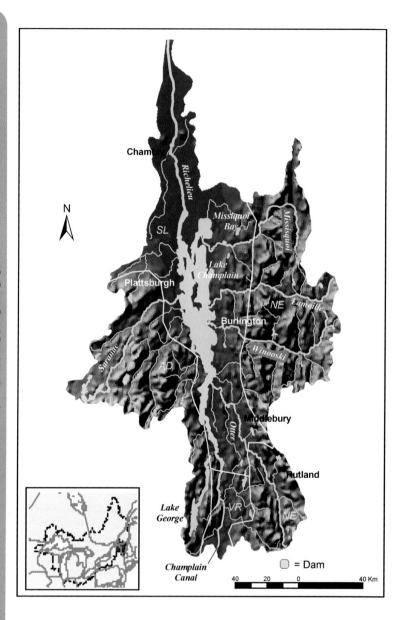

Relief: 1614 m
Basin area: 23,772 km²
Mean discharge: 341 m³/s
Mean annual precipitation: 115 cm

Mean air temperature: 4.1°C
Mean water temperature: NA
No. of fish species: 48
No. of endangered species: 0

Physiographic provinces: St. Lawrence Lowlands (SL), Adirondack (AD), New England (NE), Valley and Ridge (VR)

Major fishes: black fin shiner, largemouth bass, bowfin, banded killifish, emerald shiner, spottail shiner, river redhorse, mimic shiner, American eel, fathead minnow, fallfish, greater redhorse, yellow perch, rock bass, pumpkinseed, blunt nose minnow, brown bullhead, black crappie, white sucker

Major other aquatic vertebrates: muskrat, beaver

Major benthic insects: true flies (chironomid midges), caddisflies (Hydropsychidae, Hydroptilidae, Leptoceridae, Limnephilidae, Polycentropidae), mayflies (Baetidae, Caenidae)

Nonnative species: sea lamprey, quagga mussel, zebra mussel, Asiatic clam, water chestnut

Major riparian plants: primarily beech and maple, with spruce and fir at higher elevations

Special features: outlet for Lake Champlain; river falls ~20 m at Fryer Rapids near Chambly, Quebec

Fragmentation: dam and reservoir below Fryer Rapids

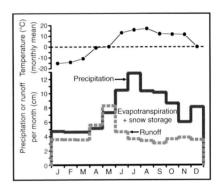

Rivière Richelieu at Saint-Jean-sur-Richelieu, Quebec (photo by G. Winkler).

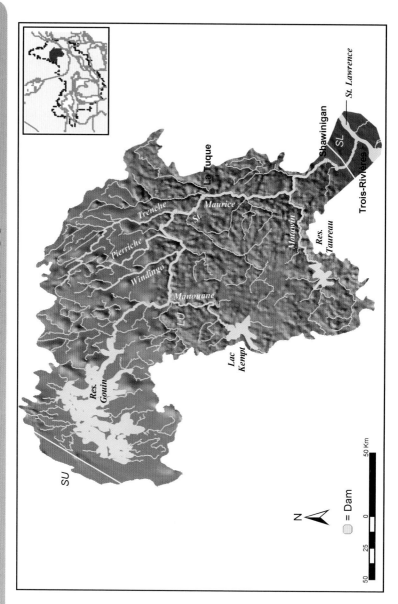

Rivière Saint-Maurice

Relief: 548 m
Basin area: 43,427 km^2
Mean discharge: 670 m^3/s
Mean annual precipitation: 93 cm

Mean air temperature: 9.7°C
Mean water temperature: 10.0°C
No. of fish species: ≥30
No. of endangered species: 5

Physiographic provinces: Laurentian Highlands (LU), St. Lawrence Lowland (SL)

Major fishes: northern pike, lake trout, lake whitefish, fallfish, pearl dace, creek chub, fathead minnow, longnose dace, shiners, brown bullhead, troutperch, rock bass, pumpkinseed, smallmouth bass, yellow perch, walleye, logperch, white sucker, longnose sucker, brook trout

Major other aquatic vertebrates: muskrat, mink, beaver, common loon, bufflehead, black scoter

Major benthic insects: NA

Nonnative species: brook trout, pearl dace, purple loosestrife

Special features: La Mauricie National Park; Mastigouche and several other provincial faunal reserves/parks in the basin

Fragmentation: 7 main-stem dams

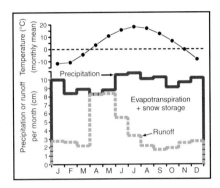

Rivière Saint-Maurice at Shawinigan, Quebec (photo by G. Winkler).

Chapter 23

Rivers of Mexico

Paul F. Hudson, Dean A. Hendrickson, Arthur C. Benke, Alejandro
Varela-Romero, Rocio Rodiles-Hernández, and Wendell L. Minckley

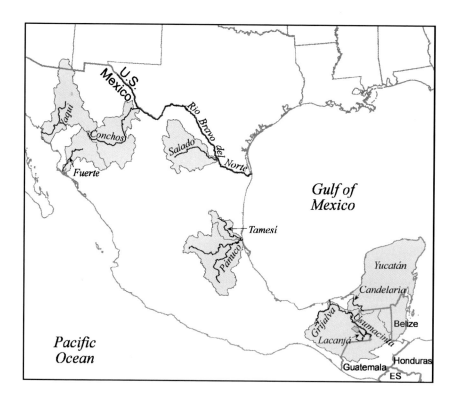

México, with an area of 1.97 million km², has approximately 150 sizable rivers, and appears to have an abundance of water resources. However, the distribution of these water resources is far from homogeneous, with northern México being extremely dry and southern México being among the wettest areas of North America. Located between 15° and 33°N latitudes, México is the warmest part of North America (see map), yet has tremendous variety in climate and topography. The combination of its mountainous topography producing strong orographic influences on precipitation and the fact that it straddles the temperate-tropical climatic divide produces a diversity of runoff patterns and river environments. With a population of 108 million (www.prb.org), or an average density of 55 people/km², tremendous pressure is placed on México's rivers for hydropower, irrigation, waste disposal, and domestic and industrial consumption.

Various accounts suggest the earliest inhabitants arrived more than 20,000 years ago and several major Mesoamerican culture regions developed large populations with sophisticated and complex cultures. Following the Olmec civilization that developed along the southeastern Mexican Gulf Coastal Plain around 4000 years ago, other major culture regions (Maya, Toltec, Huastec, and Aztec) developed within México during the Classic period (150–650 A.D.).[1] These complex civilizations were heavily dependent on México's rivers and water resources until the Spanish conquest. Far to the north, other complex cultures developed, albeit to a somewhat lesser extent and in somewhat later periods. The Mayan civilization flourished from about 600 to 900 A.D. reaching a population of approximately 5 million. They inhabited the entire Yucatán peninsula as well as the states of Chiapas and Tabasco. West of the Mayan region, the Aztecs dominated Central México from about 900 to 1521 A.D. The Aztec population had reached an estimated 25 million when conquered by the Spaniard Hernán Cortés in 1521. Cortés' conquest resulted in Spanish rule for about 300 years until Mexican independence was declared in 1810.

In this chapter, we describe 9 rivers ranging from those found in the arid north to the wet tropical climate in the south. Among these rivers, we describe two tributaries of the Río Bravo del Norte (known in the United States as the Río Grande), but the main stem is described in Chapter 5. The northern arid rivers are the Conchos, Salado, Fuerte, and Yaqui, all with mean discharges <100 m³/s, in spite of having basins >30,000 km². In contrast, the Pánuco of east-central Mexico has a mean discharge of almost 500 m³/s, and the Usumacinta-Grijalva of southern Mexico has a mean discharge of 2678 m³/s. Also described are the smaller Tamesí, and the Candelaria and Lacanja in the south.

---

1 Coe, M. D., & R. Koontz. 2002. *México: From the Olmecs to the Aztecs*, 5th Ed. Thames and Hudson, New York.

Río Pánuco

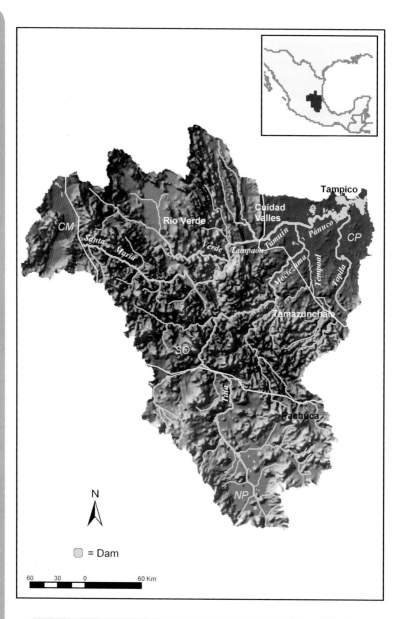

= Dam

Relief: 3800 m
Basin area: 79,100 km²
Mean discharge: 472.8 m³/s
Mean annual precipitation: 96 cm

Mean air temperature: 20°C
Mean water temperature: NA
No. of fish species: >88
(>80 native)
No. of endangered species: 7

Physiographic provinces: Neovolcanic Plateau (NP), Sierra Madre Oriental (SO), Mexican Gulf Coastal Plain (CP), Central Mesa (CP)

Major fishes: Media Luna cichlid, blackcheek cichlid, Chairel cichlid, slender cichlid, bluetail splitfin, dusky goodea, relict splitfin, jeweled splitfin, Media Luna pupfish, sheepshead swordtail, short-sword platyfish, delicate swordtail, highland swordtail, Moctezuma swordtail, Pánuco swordtail, variable platyfish, spottail chub, Pánuco minnow, bicolor minnow, chubsucker minnow, flatjaw minnow, Río Verde catfish, phantom blindcat, fleshylip buffalo, Mexican tetra

Major other aquatic vertebrates: NA

Major benthic insects: NA

Nonnative species: tilapias, grass carp, silver carp, channel catfish, blue catfish, largemouth bass, water hyacinth

Major riparian plants: NA, but see Río Tamesí

Special features: Cascada Tamul, 102 m high waterfall (at confluence of Río Gallinas and Río Santa Maria); natural springs; caves and sinkholes are common, some perhaps among world's deepest, many with abundant aquatic habitat; some of highest areas harbor cloud forests (El Cielo Biosphere Reserve)

Fragmentation: dams on Río Moctezuma, Río Tula, Río Topila; dam under construction on lower Río Tamuin

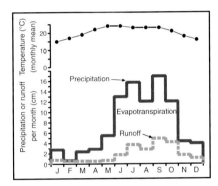

Lower Río Tampaón (Río Pánuco) (photo by P. H. Hudson).

Ríos Usumacinta–Grijalva

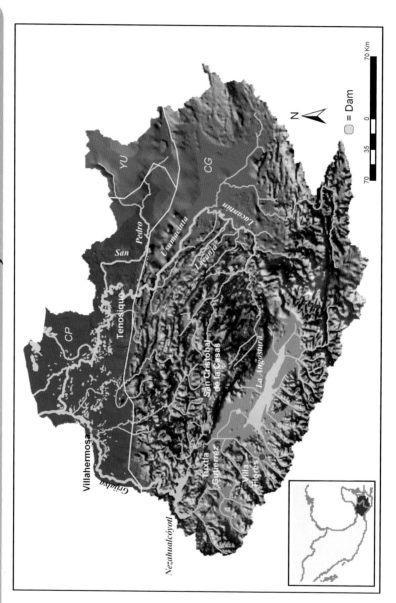

Relief: 3800 m
Basin area: 112,550 km^2
Mean discharge: 2678 m^3/s
Mean annual precipitation: 199 cm

Mean air temperature: 23°C
Mean water temperature: NA
No. of fish species: >112 (103 native)
No. of endangered species: 11

Physiographic provinces: Chiapas–
Guatemala Highlands (CG), Mexican
Gulf Coastal Plain (CP), Yucatan (YU)
Major fishes: Pénjamo tetra, headwater
killifish, Chiapas killifish, widemouth
gambusia, Chiapas swordtail, sulphur
molly, upper Grijalva livebearer,
white cichlid, Angostura cichlid, tailbar
cichlid, Petén cichlid, Montechristo
cichlid, Usumacinta cichlid, Chiapa
de Corzo cichlid, freckled cichlid,
Teapa cichlid, longfin gizzard shad,
Lacandon sea catfish, Maya sea
catfish, Maya needlefish
Major other aquatic vertebrates: river
crocodile, swamp crocodile, common
snapping turtle, tortugas blanca,
tortugas casquito, neotropical river
otter, West Indian manatee
Major benthic insects: caddisflies
(*Smicridea*, *Nectopsyche*, *Neotrichia*),
mayflies (*Camelobaetidius*,
Leptophlebia, *Traverella*, *Campsurus*)
Nonnative species: common carp,
grass carp, rainbow trout, largemouth
bass, blue tilapia, redbelly tilapia, Nile
tilapia, Mozambique tilapia

Major riparian plants: bloodwoodtree,
gregorywood, willow, button
mangrove, black mangrove, white
mangrove, American mangrove,
cattails, common reed
Special features: Pantanos de Centla
Biosphere Reserve; Laguna del Tigre
National Park in Guatemala; Selva
Maya, possibly largest remaining
tropical forest in North/Central
America; center of ancient Mayan
culture; Montes Azules Biosphere
Reserve; Lacantún Biosphere Reserve
Fragmentation: large dams on
Río Grijalva; dams proposed on
Usumacinta

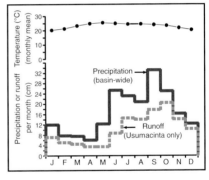

Upper portion of Río Usumacinta (Río Lacantún) (photo by H. Bahena).

Río Candelaria (Yucatán)

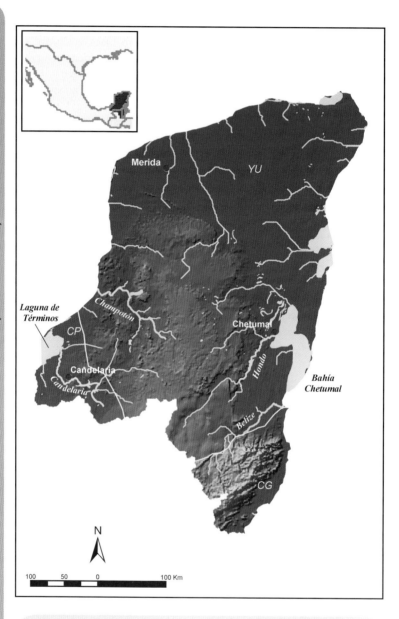

Relief: 375 m
Basin area: 10,755 km^2
Mean discharge: 46 m^3/s
Mean annual precipitation: 150 cm

Mean air temperature: 24.6°C
Mean water temperature: NA
No. of fish species: >65
(>17 freshwater, 48 estuarine)
No. of endangered species: 0

Physiographic provinces: Yucatan (YU), Mexican Gulf Coastal Plain (CP)

Major fishes: firemouth cichlid, blackgullet cichlid, yellow cichlid, yellowjacket, Mayan cichlid, yellowbelly cichlid, Montechristo cichlid, redhead cichlid, giant cichlid, stippled gambusia, Champoton gambusia, shortfin molly, picotee livebearer, pike killifish, banded tetra, Maya tetra, pale catfish, threadfin shad, bay anchovy, silver perch

Major other aquatic vertebrates: swamp crocodile, neotropical river otter

Major benthic insects: stoneflies (*Anacroneuria*)

Nonnative species: none documented

Major riparian plants: cattails, breadnut tree, bay palmetto, gumbo limbo, gregorywood, mangroves

Special features: one of the few rivers flowing through the highly karstic region of the Yucatan Peninsula; large water losses to groundwater; where ancient Mayan civilization developed

Fragmentation: no dams or reservoirs; prehistoric canal systems

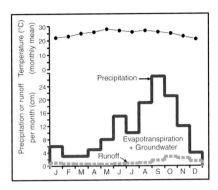

Río Candelaria in the state of Campeche (photo by A. H. Siemens).

Río Yaqui

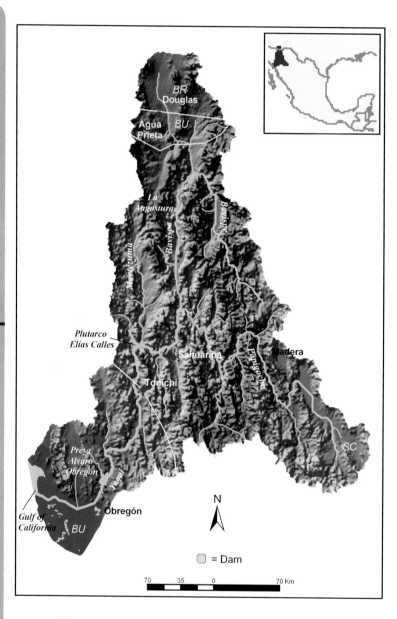

Relief: 2520 m
Basin area: 73,000 km²
Mean discharge: 78.5 m³/s
Mean annual precipitation: 48 cm

Mean air temperature: 18.4°C
Mean water temperature: 18.0°C
No. of fish species: 107
No. of endangered species: 7

Physiographic provinces: Sierra Madre Occidental (SO), Basin and Range (BR), Buried Ranges (BU)

Major fishes: Yaqui trout, Yaqui sucker, Yaqui chub, roundtail chub, Mexican stoneroller, longfin dace, Yaqui catfish, Yaqui topminnow, Leopold sucker, Cahita sucker, Pacific gizzard shad, striped mullet, striped mojarra, machete, Heller's anchovy

Major other aquatic vertebrates: water snakes (*Thamnophis* spp.), Sonora mud turtle, Tarahumara frog, Tarahumara salamander, Chiricahua leopard frog, neotropical river otter

Major benthic insects: mayflies (*Siphlonurus*, *Nixe*, *Acentrella*), stoneflies (*Capnia*, *Mesocapnia*, *Anacroneuria*)

Nonnative species: channel catfish, blue catfish, black bullhead, largemouth bass, rainbow trout, common carp, river carpsucker, green sunfish, bluegill, western mosquitofish, American bullfrog

Major riparian plants: Arizona sycamore, Arizona alder, Goodding willow, Bonpland willow, green ash, Fremont cottonwood, common reed, Chinese saltcedar, velvet mesquite

Special features: basin shared between Arizona and México (Sonora and Chihuahua); spectacular canyons (barrancas); high fish diversity for arid system, but low endemism; protected areas are Tutuaca, Papigochic, and Sierra de Ajos Bavispe

Fragmentation: 3 major dams on main stem; flow regulation over 50% of the basin

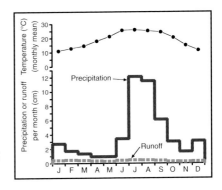

Upper Río Yaqui (photo by W. E. Doolittle).

Río Conchos

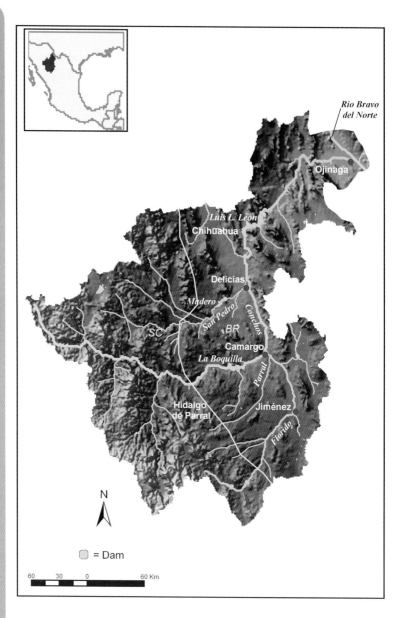

= Dam

60   30   0   60 Km

Relief: 2700 m
Basin area: 68,386 km²
Mean discharge: 20.5 m³/s
Mean annual precipitation: 48 cm

Mean air temperature: 18°C
Mean water temperature: NA
No. of fish species: 53 (>38 native)
No. of endangered species: 5

Physiographic provinces: Sierra Madre Occidental (SC), Basin and Range (BR)

Major fishes: Conchos shiner, bigscale pupfish, bighead pupfish, Salvador's pupfish, Conchos darter, yellowfin gambusia, crescent gambusia, longnose gar, largemouth bass, Mexican tetra, Río Grande cichlid, Mexican stoneroller, ornate minnow, red shiner, roundnose minnow, Tamaulipas shiner, longnose dace, Conchos chub, headwater catfish, blue catfish, channel catfish

Major other aquatic vertebrates: water snakes (*Thamnophis* spp.)

Major benthic insects: NA

Nonnative species: goldfish, common carp, warmouth, inland silverside, white bass, rainbow trout

Major riparian plants: NA

Special features: major inflow for the Río Bravo del Norte upstream of Big Bend National Park; considerable endemism of fish species and herpetofauna; specialized fauna in spring and cave habitats

Fragmentation: several dams, Presa Boquilla is largest; flow is reduced to zero for months at a time

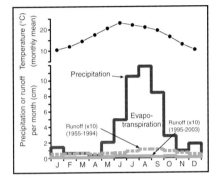

Upper incised Río Conchos (photo by W. E. Doolittle).

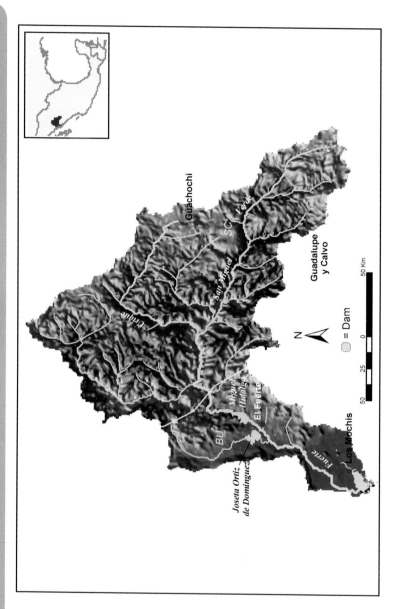

Relief: 3300 m
Basin area: 34,247 km²
Mean discharge: 31 m³/s
Mean annual precipitation: 78 cm

Mean air temperature: 24°C
Mean water temperature: NA
No. of fish species: 51
No. of endangered species: 9

Physiographic provinces: Buried Ranges (BU), Sierra Madre Occidental (SC)

Major fishes: Mexican stoneroller, ornate shiner, Mexican golden trout, diverse unisexual clones of topminnow, roundtail chub, Yaqui sucker, Yaqui catfish, mountain clingfish

Major other aquatic vertebrates: neotropical river otter

Major benthic insects: NA

Nonnative species: largemouth bass (important reservoir sport fishery), bluegill, green sunfish, rainbow trout, channel catfish, blue catfish, common carp, tilapia

Major riparian plants: NA

Special features: major tributaries descend through Barranca del Cobre

(Copper Canyon) Park; remote historic settlements; indigenous cultures (Rarámuri or Tarahumara) occupy much of the basin

Fragmentation: large dams in lower river

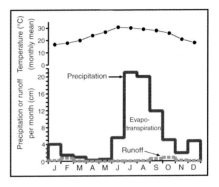

Río Fuerte, Sinaloa (photo by T. Grey).

Río Tamesí

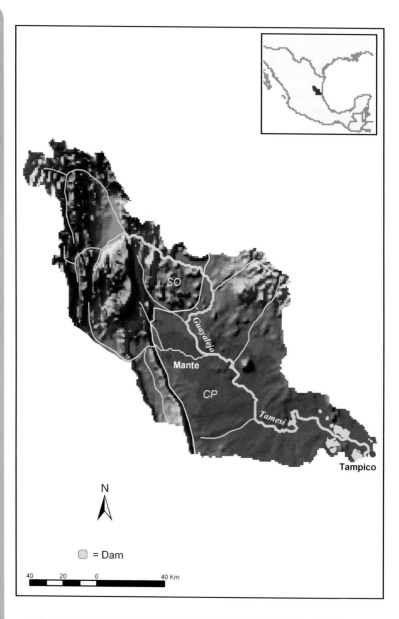

N

◯ = Dam

40   20   0   40 Km

Relief: 3353 m
Basin area: 19,127 km²
Mean discharge: 64.6 m³/s
Mean annual precipitation: 105 cm

Mean air temperature: 24°C
Mean water temperature: NA
No. of fish species: 93
No. of endangered species: 4

Physiographic provinces: Sierra Madre Oriental (SO), Mexican Gulf Coastal Plain (CP)

Major fishes: sailfin molly, Amazon molly, shortfin molly, Mexican tetra, alligator gar, variable platyfish, smallmouth buffalo, channel catfish, rainwater killifish, Río Grande cichlid, sheepshead minnow, striped mullet, hardhead catfish, phantom blindcat, chairel cichlid, slender cichlid, blackcheek cichlid, lantern minnow, Gulf gambusia, golden gambusia, spinycheek sleeper

Major other aquatic vertebrates: NA

Major benthic insects: caddisflies (*Leptomema, Nectopsyche*), mayflies (*Baetis*)

Nonnative species: common water hyacinth, hydrilla, cinnamon river shrimp, grass carp, silver carp, largemouth bass, blue tilapia

Major riparian plants: breadnut tree, gumbo limbo, cattails, mangroves

Special features: headwaters include large El Cielo Biosphere Preserve (near Ciudad Victoria); large springs at edge of karstic Sierra Guatemala feed river via caves originating at high elevations in same range with numerous aquatic cave organisms; extensive freshwater lagoon on lower 40 km of river valley

Fragmentation: main-stem dam east of Mante; several small dams

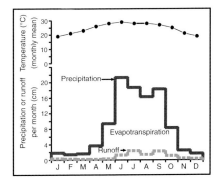

Wetlands and lagoons of the lower Río Tamesí at Tampico (photo by P. H. Hudson).

Río Salado

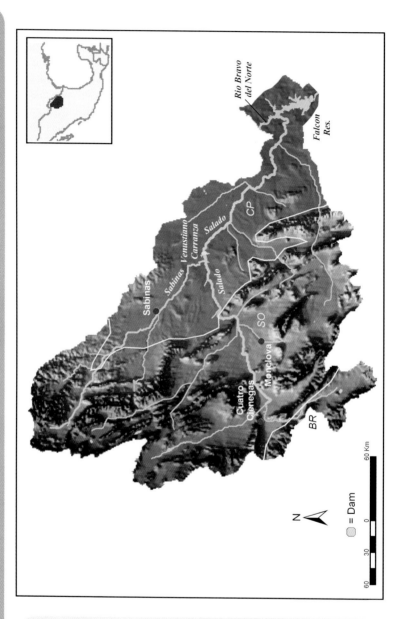

Relief: 2560 m
Basin area: 60,000 km²
Mean discharge: 10 m³/s
Mean annual precipitation: 31 cm

Mean air temperature: 21°C
Mean water temperature: NA
No. of fish species: 52
(including Cuatro Ciénegas)
No. of endangered species: 0

Physiographic provinces: Coastal Plain (CP), Sierra Madre Oriental (SO)
Major fishes: Salado shiner, tufa darter, Salado darter, Cuatro Ciénegas gambusia, robust gambusia, marbled swordtail, Cuatro Ciénegas platyfish, bolson pupfish, Cuatro Ciénegas pupfish, Mexican red shiner, Mexican tetra, largemouth bass, longear sunfish, Cuatro Ciénegas cichlid, Río Grande cichlid, roundnose minnow, Tamaulipas shiner, Devils River minnow, blue catfish, headwater catfish, channel catfish, flathead catfish, gray redhorse
Major other aquatic vertebrates: Cuatro Ciénegas box turtle, Cuatro Ciénegas red-eared slider, Cuatro Ciénegas softshell, diamondback water snake
Major benthic insects: NA
Nonnative species: warmouth, blue tilapia, spotted jewelfish, common carp, threadfin shad, red swamp crayfish, Asiatic clam, water hyacinth

Major riparian plants: mesquite, cottonwood, willows, common reed, giant reed, saltcedar, athel
Special features: major headwaters in Cuatro Ciénegas Protected Area for Fauna and Flora, a small desert valley with hundreds of large geothermal springs feeding marshes and rivers that harbor a diverse and highly endemic fauna and flora
Fragmentation: one major dam (V. Carranza) on main stem; most of main stem usually dry

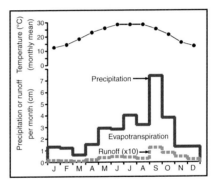

Río Salado (photo by A. Garza).

Río Lacanjá

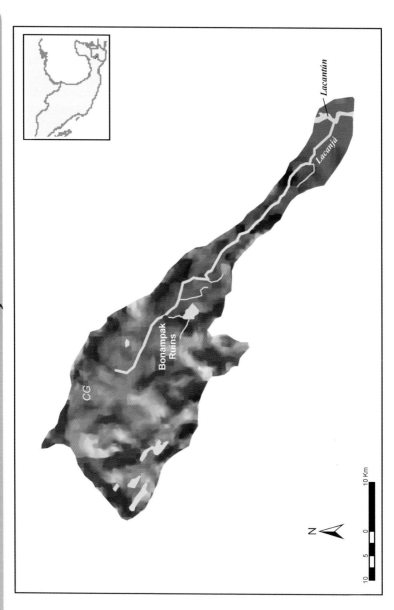

Relief: 350 m
Basin area: 800 km²
Mean discharge: NA
Mean annual precipitation: 223 cm

Mean air temperature: 25°C
Mean water temperature: 27°C
No. of fish species: 44
No. of endangered species: 6

Physiographic province: Chiapas–
Guatemala Highlands (CG)
Major fishes: Usumacinta cichlid,
white cichlid, Petén cichlid, arroyo
cichlid, freckled cichlid, bluemouth
cichlid, pantano cichlid, Palenque
cichlid, undescribed catfish, Lacandon
sea catfish, Maya sea catfish, Mexican
freshwater toadfish, white mullet,
Mexican mojarra, giant cichlid,
tropical gar, macabi tetra
Major other aquatic vertebrates: river
crocodile, neotropical river otter
Major benthic insects: NA
Nonnative species: grass carp, several
species of tilapia
Major riparian plants: willow,
cottonwood, sycamore, cypress,
*Pachira acuatica, Bravaisia
integerrima*, southern cattail, common
reed, *Thalassia, Guadua spinosa*,
waterlily

Special features: fast-flowing runs,
waterfalls with numerous runs and
rapids, floodplain clear-water lakes,
floodplain backwater, riparian
wetland; boundary between two of
most important biosphere reserves in
México, Montes Azules and Lacantún
Fragmentation: no dams; natural
fragmentation by waterfall of 15 m

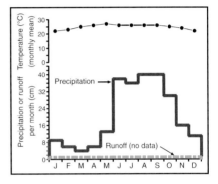

Río Lacanjá, a tributary of the Río Usumacinta (photo by H. Bahena).

Index of Rivers